W9-DGD-619

PRACTICAL
HANDBOOK

of

Curve Design
and
Generation

David von Seggern
University of Nevada
Reno, Nevada

CRC Press
Boca Raton   Ann Arbor   London   Tokyo

Library of Congress Cataloging-in-Publication Data

von Seggern, David H. (David Henry)
    Practical handbook of curve design and generation / David von Seggern
        p.   cm.
    Includes bibliographical references and index.
    ISBN 0-8493-8916-X
    1. Curves on surfaces—Handbooks, manuals, etc.   I. Title.
QA643.V665 1993
516.3'5—dc20   93-29293
                                                                            CIP

International Standard Book Number 0-8493-8916-X

Library of Congress Card Number 93-29293

Printed in the United States of America 1 2 3 4 5 6 7 8 9 0

Printed on acid-free paper

PREFACE

The publication of *CRC Handbook of Mathematical Curves and Surfaces* and its sequel, *CRC Standard Curves and Surfaces,* filled a void regarding a practical, thorough, and illustrative compendium of curves and surfaces for use in mathematics, science, engineering, and related fields. These volumes have been well received by the professionals in these fields, fulfilling an idea which this author had held for many years prior to publication.

In compiling the material for these volumes, I encountered many, many curve forms which interested me from an esthetic sense but which were deemed unbeneficial for detailed presentation to the intended audience of these volumes. However, I was keenly aware that there may be an interest in these forms among an audience composed of people in the arts, design, and graphics fields. This current volume represents an attempt to make this material accessible, understandable, and usable to this audience in the belief that they can benefit from a more thorough exposure to the mathematics of curves.

Prior to the computer era, much of the work in design was done with drafting tools or by freehand methods. These methods are still valuable and widely used; however, the computer, first a tool of science, has now become a ubiquitous tool of the arts and design. It is possible to utilize computers to generate curves and surfaces, from elementary to intricate; and such efforts are found virtually everywhere in arts and design. Even the exacting and attractive work accomplished with the drafting spline has been supplanted by programs that generate equally pleasing curves for various purposes. Thus, many people are newly engaged in using computers for art and design but perhaps have little appreciation for the mathematics behind the tools that they use.

I have deliberately refrained from making this a "computer art" book, of which there are many. The purpose is more fundamental than to supply a compendium of interesting and artful illustrations for perusal. I have endeavored to present the basic mathematics of curves in a complete and clear manner so that you can knowledgeably apply much of the material to your own work or at least appreciate the underpinnings of the graphic tools used therein. By knowing how curves are mathematically generated and how their shape is controlled, designers should be able to more fully exploit the available computer tools, modify these tools themselves, or provide input for others to modify them. It will also help to explicitly identify mathematical equations which will provide desired forms of interest. Conceivably, this book will offer inspiration to those who have used certain computer design tools for indefinite periods without really experimenting with their full capabilities.

One of the trends in the last decade has been to investigate the visual forms generated by recent accomplishments in fractals, chaos, and nonlinear dynamics. Such forms can often be quite intricate, but still pleasing to the eye. Discoveries in these computing fields are rapid and profuse, and a large body of publications already exists to show the visual results. This new arena of visual forms should be considered a valuable complement to the more fundamental and traditional arena described in this volume.

I believe that much can still be accomplished with the basic mathematical forms presented in this volume and that many appealing designs can be generated from the basic functions of mathematics: polynomials, exponentials, and harmonics. These basic functions can be understood and applied with ease.

An educational background in elementary algebra and trigonometry will suffice to be able to study, assimilate, and utilize the material in this book. Many of the essential elements of these disciplines are repeated here to reinforce an understanding of the material. Although some calculus is presented, it is of peripheral meaning; and the context should enable you to understand the terminology and usage in each case even though no formal calculus training has been taken.

By studying the material in the order of presentation, you will gain an understanding of the mathematics behind curves of all different shapes. The chapters start with fundamental polynomial, trigonometric, and exponential forms, then they apply these forms to make more and more complex shapes. The mathematical transformation of curves is then treated to give you a general approach to modifying known curves. Later chapters introduce fairly complex curves that can be composed of the fundamental building blocks of earlier chapters. I show that the piecewise curve approach is flexible and useful for very complex curves. Final chapters cover some interesting ideas in three-dimensional curves and in surfaces. I have approached these three-dimensional forms as extensions of two-dimensional curves to make a solid connection with the main body of this volume and to enable you to more easily visualize the three-dimensional forms.

Some remarks on the numerous figures are appropriate. I have deliberately tried to keep each figure simple to clearly illustrate a single point. Many figures are presented without axes or scales; and in such cases, you can assume that the aspect ratio (vertical scale to horizontal scale) is unity. Annotated axes are present whenever there could be doubt about the scales, the limits of the graphs, or the aspect ratio. Figures for three-dimensional curves or surfaces have been augmented with a "box" that surrounds the graphical object and provides a clear indication of the orientation of the object.

I believe it is important to add that this book was almost entirely an electronic product of the *Mathematica*® program of Wolfram Research, Inc. An initial draft of the text was prepared simultaneously with the figures using *Mathematica*.® The ability to compose both at the same time within the same document is a powerful feature which enables one to quite effortlessly and productively complete a project of this scope. The text was later copied to a standard word processor for final revisions, review, and enhancements. The more complex equations were set with the *Expressionist* program. Finally, the *Mathematica*® figures were converted to Encapsulated PostScript (EPS) files prior to submission.

David von Seggern
University of Nevada
Reno, Nevada 1993

THE AUTHOR

David H. von Seggern, Ph.D., is a seismologist currently at the Seismological Laboratory, University of Nevada-Reno. He is the Seismic Network Manager for the university studies at Yucca Mountain, Nevada, the site of a proposed underground nuclear waste repository.

He previously worked for Teledyne Geotech in Alexandria, Virginia, on numerous aspects of underground-nuclear-test detection. During that time, he authored or co-authored numerous professional papers and company reports on the subject. He completed his education at the Pennsylvania State University with a dissertation on earthquake prediction which included an early application of fractal theory in seismology.

Later, at Phillips Petroleum Company, Dr. von Seggern specialized in applying computer graphics to the problems of processing and interpreting seismic data, promoted seismic modeling as an aid in data interpretation, and did research in seismic imaging methods using supercomputer technology.

TABLE OF CONTENTS

Chapter 1

INTRODUCTION TO MATHEMATICAL FUNCTIONS

1.1. COORDINATE SYSTEMS

1.1.1. Axes

Because this book is basically concerned with the visual aspects of mathematical functions, it is appropriate at the beginning to introduce the visual backdrop for these functions. This backdrop is a coordinate system (there are many) with *axes* along which the variables of the functions take on values. This will be apparent as the particular systems are described. For two-dimensional functions, the axes lie on an infinite flat two-dimensional plane. These axes can be considered as rulers along which one marks the distances of the variables from a point called the *origin*, where both variables take the value zero. The axes may be straight or curved although straight axes are most common. Furthermore, the ruling on an axes will most often be linear (as almost always seen on common rulers) whereby equal intervals of the variable are given equal spacing on the axis. Sometimes axes can be nonlinear (for example, logarithmic); however, this book will utilize, with only a few exceptions, linear axes throughout.

1.1.2. Cartesian Coordinates

By far the most common coordinate system is the *Cartesian* system, named after the French mathematician and philosopher Descartes. This is true because it is also the most natural and easy to understand. It consists of two straight, linear axes joined at right angles at the origin as shown in Figure 1.1.1. The horizontal axis is usually labeled the x axis, thus measuring the x variable; and the vertical axis is usually associated with the y variable. By convention, the positive numbers are to the right on the x axis and upwards on the y axis. Each coordinate can range from negative infinity to positive infinity. There is no reason why the axes cannot be associated with other named variables, such as u and v or s and t. The variables will be related by some equation that provides a link between the two values for a particular point, given as (x, y), on a curve in this coordinate plane. The point (x, y) is called a *coordinate pair*. The system shown here is for a two-dimensional plane, but the Cartesian system naturally extends to three-dimensional space when an additional axis is placed perpendicular to the other two. The third axis is usually called the z axis.

1.1.3. Polar Coordinates

Another useful and common coordinate system for two related variables is the *polar* system. This system combines a radial coordinate, which is the distance from the origin (disregarding direction from the origin) with another coordinate that specifies the angle at which the distance is measured. This coordinate system is illustrated in Figure 1.1.2. By convention, the angle θ is measured in a counterclockwise sense from a horizontal line extending rightward from the origin. While the value of r

1

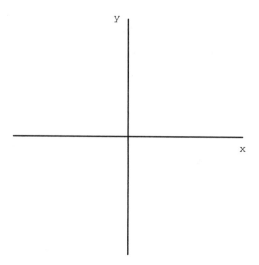

FIGURE 1.1.1. Cartesian coordinates.

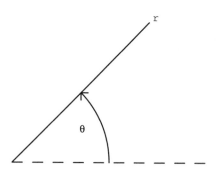

FIGURE 1.1.2. Polar coordinates.

may be from negative to positive infinity, the value of θ only ranges from 0 to 360 degrees. More commonly, the angle is expressed in radians; and, because there are 2π radians per 360 degrees, the range of θ is from 0 to 2π radians. (Recall that the circumference of a circle equals 2π times its radius.) The coordinate pair for a point in the polar system is then (r, θ). In considering a point with negative r for a given θ, one sees that the same point can be specified by positive r with θ increased by π radians (exactly half the way around the circle). Thus, each point in the plane can have two coordinate pairs: (r, θ) and $(-r, \theta + \pi)$. This phenomenon is not shared by the Cartesian coordinate system which has a unique coordinate pair for each possible point in the plane.

A simple relation exists between the polar coordinates and the Cartesian ones. It comes from basic trigonometry which provides the two equations:

$$x = r\cos(\theta)$$
$$y = r\sin(\theta)$$

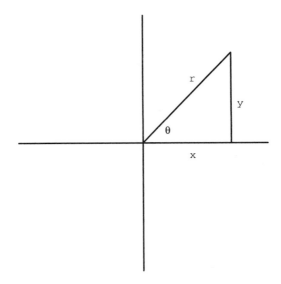

FIGURE 1.1.3. Cartesian coordinates from polar coordinates.

Figure 1.1.3 illustrates this relation: similar to that used to define the Cartesian coordinates. Note that the sides x and y form a right triangle with r being the hypotenuse. The ability to think and visualize in terms of Cartesian coordinates is so natural that, even when curves are computed using polar coordinates, the curve will often be presented on a background of the Cartesian axes and labeled accordingly. However, the mathematical expression of the curve in polar coordinates may be much simpler, in many cases, than the corresponding Cartesian expression; and it also gives much readier insight into the form of the curve.

1.2. NATURE OF FUNCTIONS

The coordinate planes described above contain an infinity of points or coordinate pairs. This book focuses on curves that are related subsets of points in this plane. These subsets are usually connected to form a *continuous* curve. Continuity is a mathematical concept that can be treated with great rigor; however, here we rely on a visual sense to determine whether a curve is continuous. Note that the term "curve" can apply to a straight line, which is simply a special case of a curve.

The subset of points that form a mathematical curve are related by a mathematical *equation* whose right-hand side is a *function*. In the case of Cartesian coordinates, a function is expressed *explicitly* as $y = f(x)$, which one reads as "y equals a function of x." Here, $f(x)$ stands for some mathematical expression that has only the variable x appearing in it with everything else constants. The following are examples of explicit equations involving functions of x:

$$y = a - bx + cx^3$$
$$y = 4\cos(2\pi x) + \sin(5\pi x)$$

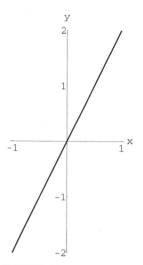

FIGURE 1.2.1. Graph of 2x.

$$y = [x^2 - 1]^{1/2} + k$$
$$y = a \exp(bx^2)$$

Note that no x appears on the left-hand side of the equation. The variable on the left-hand side is called the *dependent* variable, and the variable on the right-hand side is called the *independent* variable. The exact value of y "depends" on the numerical value assigned to x. Along with the independent variable, there are always one or more *constants* that appear; constants are assumed to have particular values and not to vary during evaluation of the function when x is varying. They may have numerical values (for example, 5) or literal values (for example, k). Clearly, one cannot evaluate an expression to a number until literal values are assigned numerical values.

A simple example of an explicit function is $y = 2x$: which one could read as "y equals two times x." This particular function (more particularly, the set of points defined by this function) is plotted in Figure 1.2.1. We call the actual plotted line of a function the *graph* of the function. In this plot, the x axis extends from -1 to $+1$; in general, we call the span of real numbers over which a function is evaluated the *domain*. The dependent y variable also has a span which is governed by the x values; we call the span of the associated y values the *range*. For this plot, as for most others, it is understood that the set of points defined by the equation actually extends beyond the domain and range given by the markings on the axes. It is common practice to plot equations only over a domain of x sufficient to completely show the character of the function. Clearly, the simple example of Figure 1.2.1 could be plotted over any domain of x and still appear relatively the same.

Some examples of equations in polar coordinates are

$$r = \cos(\theta)$$
$$r = a(1 + 2\theta)$$

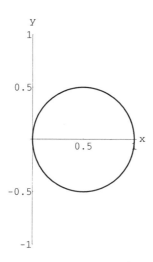

FIGURE 1.2.2. Graph of $r = \cos(\theta)$.

$$r = 1 + \sin(\theta)\cos(\theta)$$
$$r = a\exp(b\theta^2)$$

Consider the graph of the first equation, $r = \cos(\theta)$, as shown in Figure 1.2.2. Although this is a polar equation, it is plotted on the background of Cartesian axes. This equation is somewhat unusual in several respects: 1) the domain of θ required to make a complete figure is only 0 to π, halfway around a full turn of the polar axis; 2) the value of the dependent variable r ranges from -1 to 1 (recall that the cosine function is equal to $+1, 0, -1, 0$ at $0, \pi/2, \pi, 3\pi/2$, respectively) requiring a "reflection" of half of the points through the origin to values of θ which are at $\theta + \pi$; and 3) the graph of this equation is a *closed* curve with all points connected.

Thus far, equations have all been written with the dependent variable, y or r, alone on the left-hand side of the equation with the independent variable on the other side. This form can be generalized as $y = f(x)$ or as $r = f(\theta)$ and is called the *explicit* form of a function. Another form sometimes used is generalized as $f(x, y) = 0$ and is called the *implicit* form. Some examples of implicit functions should clarify what this means:

$$xy^2 - 5 = 0$$
$$ax + by + c = 0$$
$$r^2\cos^2(\theta) - 1 = 0$$
$$r(1 + 2\sin\theta) - r^3 = 0$$

For many equations, the implicit form will appear to be much simpler than the explicit form. In order to plot curves though, we need the explicit form for efficient calculations. To obtain the explicit form from the implicit one, some manipulation is required; this may be difficult or even impossible in many cases. For these reasons, this text avoids dealing with implicit forms of equations.

A final functional form we often encounter is the *parametric* form. In this form, both x and y are made to be dependent on a third, independent parameter, say t. The graph of the function is still made from the x and y coordinates: each evaluated for a given range of the parameter t. Here are a few examples of parametric functions:

$$x = \cos(t); \quad y = 1 + t^2$$
$$x = 2t + t^3; \quad y = t^2 + 6t^4$$
$$x = e^t - 1; \quad y = \sin(2\pi t)$$

Parametric functions are generally able to produce more complex graphs than explicit or implicit forms and are especially useful when we want to exactly control the x or y component of the curve.

Chapter 2

POLYNOMIALS

2.1. ORDINARY POLYNOMIALS

2.1.1. Form of the Ordinary Polynomial

We have defined a function in the $x - y$ plane simply as y (the dependent variable) equals some function involving x (the independent variable) or $y = f(x)$. Let us define terms, factors, and exponents within functions. A *factor* is a variable or constant that appears within a *term*, which are the entities separated by plus (or minus) signs within a functional expression. For instance, the following are terms, each with two or more factors, clearly separated by spaces:

$$5x$$
$$3.87ax^3$$
$$2\cos(t)$$
$$k\log(t)$$

Here, we introduced one factor within a term, specifically x^3, having an *exponent*. The exponent is a shorthand way of indicating multiplicity of identical factors. In this case, x^3 means (xxx) and is usually called "x to the third power" or just "x to the third."

For its expression, the *ordinary polynomial* has a summation of terms of the form cx^n where c is a constant. In general, we write the ordinary polynomial as

$$y = a_0 + a_1x + a_2x^2 + \cdots + a_nx^n$$

where the a_i are called the *coefficients* (any of which may be zero); and all exponents are understood to be whole numbers greater than zero. The following are examples of ordinary polynomial functions:

$$2x - 3x^4$$
$$5 + x^2 - 7x^3$$
$$1.5682 + 5.08x^3$$

Before looking at polynomials, we should examine the graphs of the individual terms of form x^n called *monomials*. The monomials have special names for the smaller values of n:

$$n = 1 \text{ linear}$$
$$n = 2 \text{ quadratic}$$
$$n = 3 \text{ cubic}$$
$$n = 4 \text{ quartic}$$
$$n = 5 \text{ quintic}$$
$$n = 6 \text{ sextic}$$

7

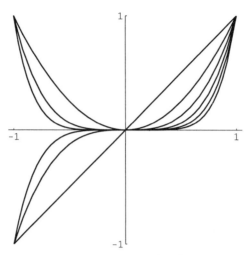

FIGURE 2.1.1.   Graphs of the monomials x^1 to x^6 in the domain $[-1, 1]$.

A function having several of these terms takes on the name associated with the term having the highest n. Figure 2.1.1 shows the graph of $y = x^n$ with the exponents n equal to one through six for x in the domain $[-1, 1]$. Note that the curves have increasingly larger "bend" as the exponent is increased. Also note that those curves with exponents of 1, 3, and 5 change sign for negative x. Whereas the curves for positive exponents are reflected from the positive x domain about the y axis; those for negative exponents are futher reflected about the x axis. We call *even* those functions which reflect once and *odd* those functions which reflect twice. The relations between the plotted portions in the negative and positive domains of x are:

$$f(-x) = f(x) \quad \text{even functions}$$
$$f(-x) = -f(x) \quad \text{odd functions}$$

Such functions are also called *symmetric* and *antisymmetric*, respectively, about the y axis. Again using the basic term x^n, the plot in Figure 2.1.2 is generated when the domain of x is allowed to enlarge to $[0, 10]$. Note the increasing slope of the graph beyond $x = 1$ as n increases.

2.1.2.   Roots and Critical Points
Of all the points in the graph of polynomials, there are certain points of special character. The first set of points we will consider are called the *roots*, which are those x points where the function equals zero. (This definition should make it apparent why the roots are also called the *zeroes* of the function.) We express this as

$$f(x) = 0$$

and find all real x that solve this equation. We qualify these solutions with "real" because basic algebra tells us that some of the roots may be imaginary numbers (or complex numbers if we allow the coefficients of the polynomial to be imaginary or

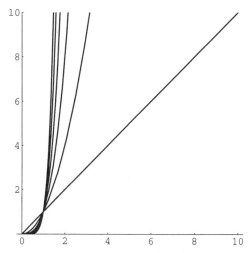

FIGURE 2.1.2. Graphs of the monomials x^1 to x^6 in the domain [0, 10].

complex numbers). Only the real roots are of interest in regard to the graph of the function. There are as many roots (real or otherwise) to a polynomial function as the largest exponent among all the terms. Only by actually solving the above equation can we determine how many roots are real and how many are imaginary. Also, some roots may be multiple.

Because each distinct real root implies a "zero crossing" or at least a "zero tangency" of the graph of the function, the shape of the curve is somewhat predictable on the basis of knowledge of the roots. Let us form a polynomial by algebraic multiplication of *binomial* factors of the form $(a+bx)$. This way, we know the factors of the polynomial; and each factor gives a root because setting any of the factors equal to zero satisfies the above equation. Each factor gives a root of the form $x = -a/b$. For instance, take the following factors:

$$1 + 2x$$
$$5 - 8x$$
$$1 - 10x$$

Multiplied together, they give the function $5 - 48x - 36x^2 + 160x^3$ plotted in Figure 2.1.3. Note that the roots appear exactly at the points given by each factor when it is set to zero and solved for $x(-1/2, 1/10, 5/8)$.

If a function is composed of more than one identical factor, the roots will be identical. Consider the factor $(1 + x)$. By multiplying it with itself once, we obtain a quadratic function. By multiplying it with itself once again, we obtain a cubic function. The former and latter functions are plotted in Figures 2.1.4 and 2.1.5, respectively. Note the coincidence of the roots at one value of $x(x = -1)$.

We stated above that some roots may be imaginary or complex numbers. (You should know that the imaginary number i is equal to $\sqrt{-1}$ and that consequently $i^2 = -1$.) It is proven in basic algebra that if the function itself is entirely real (no

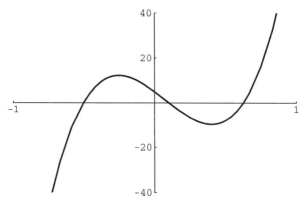

FIGURE 2.1.3. Graph of $5 - 48x - 36x^2 + 160x^3$.

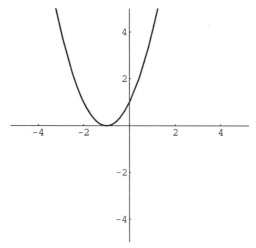

FIGURE 2.1.4. Graph of $1 + 2x + x^2$.

imaginary or complex coefficients) any imaginary roots must occur in pairs: $x = ia$ and $x = -ia$. If they are in pairs like this, then the product will be $x^2 + a^2$ because $i^2 = -1$; thus, the result is real. By specifying one or more pairs of imaginary roots, along with real roots, a real function is constructed.

Let us define the change of a function by $df(x)$ and the corresponding change in x by dx. The *derivative* or *slope* of the function is then given by $df(x)/dx$; visually, the slope is associated with the line tangent to the curve when it is graphed. In addition to roots, the other points of interest in polynomials are where the slope of the graph becomes zero. These points are termed *critical* points, of which there are three kinds: *minimum, maximum,* or *inflection* points.

The function defining the slope is equal to the first derivative of the function itself; it is written in shorthand as $f'(x)$. Determining $f'(x)$ for general functions is a topic of calculus; however, for polynomials, it is simple: for each term of the polynomial, multiply the constant by the exponent and then reduce the exponent by one. These

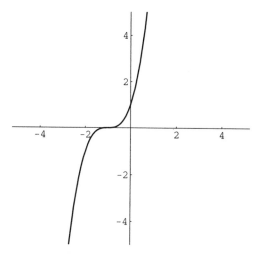

FIGURE 2.1.5. Graph of $1 + 3x + 3x^2 + x^3$.

examples should clarify the differentiation of polynomials:

$$f(x) = 4x + 2x^3; \qquad f'(x) = 4 + 6x^2$$
$$f(x) = 10x^2 - 7x^4 + x^6 \quad f'(x) = 20x - 28x^3 + 6x^5$$

By setting $f'(x) = 0$ and solving for its roots, we find the x points where the slope is zero and thus the critical points. The change of slope is associated with the second derivative as $df'(x)/dx$ or in shorthand as $f''(x)$. By evaluating $f''(x)$ at the critical points, we determine what kind they are. If $f''(x) > 0$, the critical point is a minimum; if $f''(x) < 0$, it is a maximum; and, if $f''(x) = 0$, it is an inflection. The plot of the quadratic function in Figure 2.1.4 exhibits a minimum point, and the plot of the cubic function in Figure 2.1.5 exhibits an inflection point. For these simple functions, the critical points are also roots of the function. This is seldom the case for general polynomials.

2.1.3. Behavior of the Graph of the Ordinary Polynomial

Because the ordinary polynomial is a sum of monomial terms (with perhaps constants as multipliers in each term), the effect of each term is additive in the whole expression. At a particular value of x, some terms will tend to dominate while others are relatively unimportant in contributing to y. For x general, it is not possible to state which terms are important and which are not. However, in the limits of $x \to 0$ and $x \to \infty$, it is possible to do so. These are often the regions of x that are of interest.

First consider when $x \to \infty$. As seen in Figures 2.1.4 and 2.1.5, it is clear that due to the exponent n, which raises x to a power, the term with the highest exponent will dominate irregardless of the value of the constant coefficient. This can be stated as

$$y \sim x^n \quad (n \text{ largest of all exponents})$$

This approximation can be illustrated by the following equation as an example:

$$y = 4x + 7x^3 + x^5$$

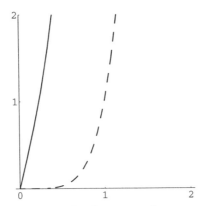

FIGURE 2.1.6. Graphs of $4x + 7x^3 + x^5$ (solid) and x^5 (dashed) in the domain [0, 1].

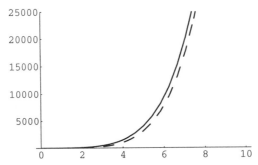

FIGURE 2.1.7. Graphs of $4x + 7x^3 + x^5$ (solid) and x^5 (dashed) in the domain [0, 10].

Figures 2.1.6 and 2.1.7 show the full function as a solid line and the approximation, x^5, as a dashed line. (Note the difference in x scale between the two figures.) We see that for small x the curves are very different; however, for large x, there is little visual difference between the curves even though the coefficient on x^5 is smaller than the coefficients on either of the other two terms. This example shows why the polynomial is basically controlled by the term with the largest exponent for large x. If that exponent is even, the curve goes to infinity in the same direction for both large negative and large positive x. This direction depends on the sign of the term in which the largest exponent occurs. If that exponent is odd, the curve goes to infinity in opposite directions; and the directions for positive and negative x depend on the sign of the term again.

In contrast, now consider the behavior of the same function for small $x(x \ll 1)$. Here, we find that the function is well approximated by the term with the smallest exponent. To illustrate this, examine Figure 2.1.8, which shows the same function for a small interval very near zero. Figure 2.1.6 indicated that the term $4x$ would be a poor approximation over the whole interval of zero to one; however, Figure 2.1.8 shows that it is an excellent approximation for very small x.

Therefore, ordinary polynomial expressions can be approximated by only one of their terms in either of two domains: x very large or x very small. In the former,

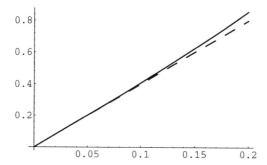

FIGURE 2.1.8. Graphs of $4x + 7x^3 + x^5$ (solid) and $4x$ (dashed).

the approximation is simply the term with the largest exponent; and, in the latter, it is simply the term with the smallest exponent (disregarding the constant a_0). These conclusions are true regardless of the magnitude of the constants that multiply the terms. Even large constant factors on terms with lower exponents fail to affect the appearance of the curve for large values of x. Similarly, terms with large exponents are unable to have any significant affect near $x = 0$ even if their constant factors are large.

What does the constant a_0 do to the appearance of a curve? The answer is that it has no affect on it whatsoever other than to move it up or down the y axis. The effect of a_0 in a polynomial equation is to translate the curve, unchanged, in the vertical direction. How do we achieve a similar translation horizontally? By replacing every occurence of x in the function with $x - a$, the curve would be translated a units to the right; if the replacement is $x + a$, the translation would be to the left. Chapter 5 will treat translations in a more rigorous way, but it is important to remember how basic translations are accomplished. These effects apply not just to polynomial functions but to all functions in general.

We have described the behavior of a polynomial for large and small x. But for intermediate x, we claim that the behavior can also be well quantified with some simple analysis. Recall that we can get the roots of the function by setting $f(x) = 0$ and solving this equation for real values of x. These values then mark where the curve crosses or touches the x axis. Further recall that the critical points can be found by setting $f'(x) = 0$ and solving it for real values of x. We will find that a minimum or maximum will lie between each adjacent pair of roots and that inflection points may lie anywhere. Whenever these points are found, the function $f(x)$ can be evaluated to get the full coordinate pair (x, y). We will then find that the full graph of the curve is fairly well constrained by the roots, the critical points, and the small-x and large-x behavior.

In spite of this quantification, it is not easy to infer the behavior of an arbitrary polynomial function just from the written function itself. To make this point, try to imagine the behavior of $0.6x - 0.5x^2 - 0.3x^3 + 0.1x^4$ in the domain of $[-4, 4]$. We know that it has both even and odd terms; therefore, it should have no particular symmetry. Also, we know that the behavior for large x is associated with the x^4 term: which is positive. Consequently, the curve goes towards infinity in a positive sense on both sides of the y axis. Beyond this, the imagination fails to see the curve. It

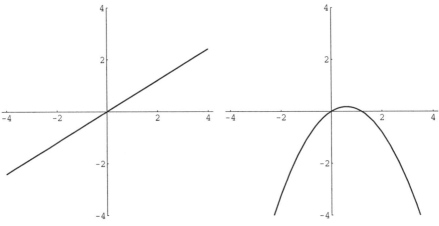

FIGURE 2.1.9. Graph of $0.6x$.   **FIGURE 2.1.10.** Graph of $0.6x - 0.5x^2$.

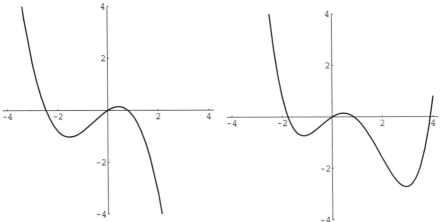

FIGURE 2.1.11. Graph of $0.6x - 0.5x^2 - 0.3x^3$.   **FIGURE 2.1.12.** Graph of $0.6x - 0.5x^2 - 0.3x^3 + 0.1x^4$.

requires algebraic analysis to produce the roots and critical points, or it requires a full graphing of the function. Because it is difficult to visualize as a whole, this function is "built up" in several sequential plots shown in Figures 2.1.9 to 2.1.12. We do this by adding one term at a time and graphing the result after each addition starting with the linear term x up through the term having x^4. Note that for the largest values of x (negative and positive), the curve is governed by the term with the highest exponent. For the smaller values of x though, the shape of the curve changes only slightly with the addition of terms having larger and larger exponents. In the vicinity of $x = 0$, the first linear term's behavior is evident even after all the other terms are added.

2.1.4.   Even Polynomial Functions
Of interest will be those polynomial functions that contain only even or only odd monomials because they will retain their symmetry, either even or odd, regardless of

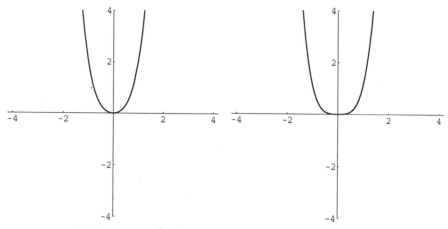

FIGURE 2.1.13. Graph of $x^2 + x^4$. **FIGURE 2.1.14.** Graph of $0.1x^2 + x^4$.

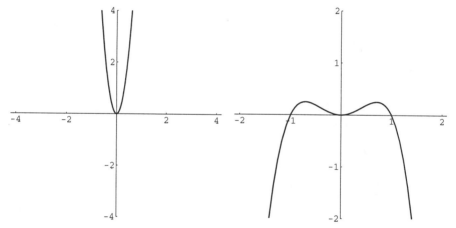

FIGURE 2.1.15. Graph of $10x^2 + x^4$. **FIGURE 2.1.16.** Graph of $x^2 - x^4$.

the number of such terms. Consider the even function $y = x^2 + x^4$ plotted in Figure 2.1.13.

Now let's consider some curve design strategy. We know that the curve near $x = 0$ is controlled by the x^2 term. By adjusting the constant on this term, we can flatten or exaggerate the concavity near $x = 0$. Compare the two graphs in Figures 2.1.14 and 2.1.15 having $0.1x^2$ and $10x^2$, respectively, with that of Figure 2.1.13 that has $1.0x^2$.

What if the sign of the second term is changed to negative? The graph changes considerably to that shown in Figure 2.1.16. Recall that the region of the curve near $x = 0$ is dominated by the term with the smallest exponent, x^2 in this case. The quadratic term with positive sign will be concave upward; this effect is seen in the graph around $x = 0$. However, for larger x, the graph is dominated by x^4; because its sign is negative, it is concave downwards. In the middle region of x, the two terms interact to produce the points of zero slope on either side of the y axis. These points

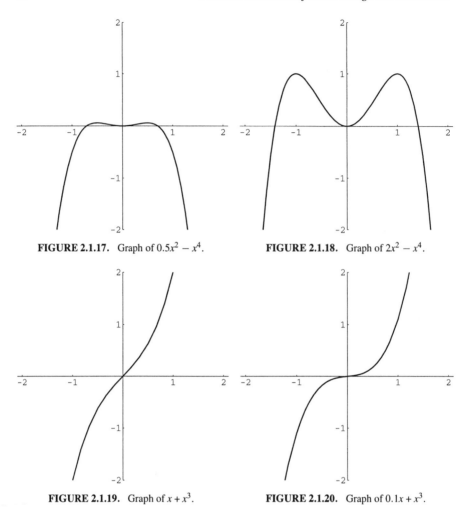

FIGURE 2.1.17. Graph of $0.5x^2 - x^4$.

FIGURE 2.1.18. Graph of $2x^2 - x^4$.

FIGURE 2.1.19. Graph of $x + x^3$.

FIGURE 2.1.20. Graph of $0.1x + x^3$.

are determined exactly by setting $f'(x) = 0$ which has the roots $0, -.707, +.707$. The first point is a mimimum and the other two are maxima. Again, we can tailor the curve near $x = 0$ by varying the constant associated with the x^2 term. This is shown in Figure 2.1.17 for $0.5x^2$ and in Figure 2.1.18 for $2.0x^2$.

2.1.5. Odd Polynomial Functions

Polynomials consisting of terms with powers of x that are only odd are also of special interest. These functions retain odd symmetry about the y axis. Consider the odd function $y = x + x^3$ plotted in Figure 2.1.19.

This curve has only one real root and no critical points. As for even-power polynomials, we can do some basic curve design by varying the constant on the linear x term to shape the curve near $x = 0$. This is illustrated in Figure 2.1.20, which uses $0.1x$, and in Figure 2.1.21, which uses $3.0x$.

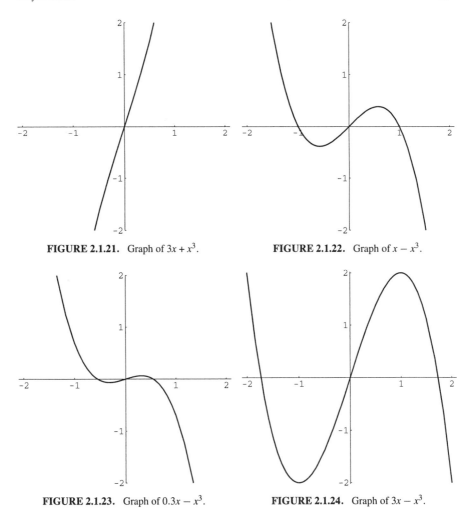

FIGURE 2.1.21. Graph of $3x + x^3$.

FIGURE 2.1.22. Graph of $x - x^3$.

FIGURE 2.1.23. Graph of $0.3x - x^3$.

FIGURE 2.1.24. Graph of $3x - x^3$.

Let us change the sign on the x^3 term. The function $x - x^3$ has three roots: $0, -1$, and $+1$. The derivative of the function is $1 - 3x^2$; this has roots at $x = -0.577$ and $+0.577$ that are critical points: in this case, a minimum and a maximum. The graph of this function appears in Figure 2.1.22. Again, let us look at the adjustments in shape that can be made by varying the constant on the linear term. Figure 2.1.23 shows $0.3x$ and Figure 2.1.24 shows $3.0x$.

2.1.6. Factors of x^n

For many polynomials, x^n can be factored out where n is arbitrary. This implies that the function has a multiple (n times) root at $x = 0$. The curve necessarily passes through zero. If the remaining polynomial is symmetric (even powers), then the effect of the x^n factor will be either to preserve this symmetry ($n =$ even integer) or to convert it to antisymmetry ($n =$ odd integer). If the remaining polynomial is antisymmetric,

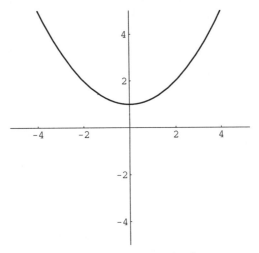

FIGURE 2.1.25. Graph of $1 + x^2/4$.

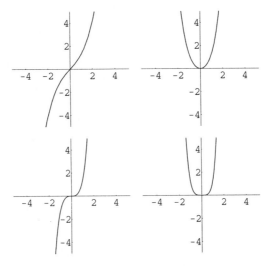

FIGURE 2.1.26. Graphs of $x^n(1 + x^2/4)$ for $n = 1, 2, 3, 4$.

the x^n factor either preserves this property if n is even or converts it to symmetry if n is odd.

The facility offered by the simple factor x is often desirable in designing a curve for it always changes the symmetry of a function which it multiplies. However, the linear x factor carries the additional weigh of enhancing the amplitudes at large x relative to those at small x. Factors with larger exponent n will weigh the values at large x even more. We can illustrate the effect of x^n factors with the parabola $1 + x^2/4$ plotted in Figure 2.1.25. The effects of multiplying this function by the factors x^n with $n = 1, 2, 3$, and 4 are shown in Figure 2.1.26.

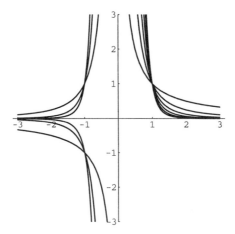

FIGURE 2.2.1. Graphs of $1/x^n$ for $n = 1, 2, 3, 4, 5, 6$.

2.2. RATIONAL POLYNOMIALS

2.2.1. Form of the Rational Polynomial

As defined in Section 2.1.1, the *rational polynomial* is a quotient of two polynomials. The general form is

$$y = \frac{b_0 + b_1 x + b_2 x^2 + \cdots + b_m x^m}{a_0 + a_1 x + a_2 x^2 + \cdots + a_n x^n}$$

Either the numerator or the denominator can be monomial. The numerator can also simply be a constant. This is not true for the denominator because the expression could then be rewritten simply as an ordinary polynomial. We will see that the rational polynomial allows considerably more flexibility in the shapes of curves that can be generated when compared with the ordinary polynomial. The denominator also introduces an important new complexity. Consider the above equation as $f(x) = g(x)/h(x)$. Recall the discussion of roots from Section 2.1.2. The real roots, or zeroes, of $h(x)$ will cause the rational polynomial to evaluate as infinity. The x values of these roots of $h(x)$ are called *poles* of the rational function; they clearly have a marked effect on the appearance of the graph of the function—even more so than the zeroes of the numerator $g(x)$.

The variety of graphs obtainable with rational polynomials is large; but it is worthwhile to study some of the simpler forms that show either symmetry or antisymmetry.

2.2.2. The Special Form c/x^n

If the numerator is simply a constant and the denominator is a monomial, the general form in Section 2.2.1 reduces to c/x^n. This form has a pole at $x = 0$ with a symmetric or antisymmetric graph depending on whether n is even or odd, respectively. Figure 2.2.1 shows the graphs of this function for n equal to one through six. Note that the bend in the curve becomes greater as n is increased.

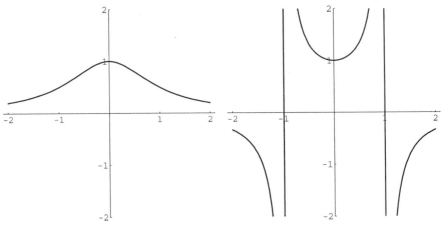

FIGURE 2.2.2. Graph of $1/(1 + x^2)$.   **FIGURE 2.2.3.** Graph of $1/(1 - x^2)$.

2.2.3.   Even Rational Polynomials
The simplest form of an even rational polynomial is

$$y = \frac{b}{a^2 + x^2}$$

or

$$y = \frac{b}{a^2 - x^2}$$

The first has no real zeros in the function for the denominator and thus no poles; the second has poles at $x = -a$ and $+a$, which are the zeros of the denominator. In spite of there being only a sign change between these equations, the graphs are radically different as shown in Figures 2.2.2 and 2.2.3. In Figure 2.2.3 the vertical lines are not part of the actual function. They lie at the poles and represent the discontinuity where the function goes to infinity in one direction and returns from infinity in the other direction. This is an *infinite discontinuity* and is the rule for rational polynomials; but we will find that discontinuities may be finite for other types of functions.

By changing the constant a, one can easily manipulate the shape of the function with $a^2 + x^2$ in the denominator. The smaller the value of a, the larger the function becomes when $x = 0$ as illustrated in Figure 2.2.4. The curve approaches the same value for large x, regardless of a.

The effect of changing a in the function with $a^2 - x^2$ in the denominator is to move the poles of the function to the new value of a: positive and negative. Correspondingly, the value at $x = 0$ moves up or down the vertical axis in a manner identical to changing a in the function with $a^2 + x^2$ in the denominator.

Next consider the functions where quartic terms a^4 and x^4 replace the quadratic ones in the above equations. Figures 2.2.5 and 2.2.6 show the graphs respectively. We can see that increasing the exponents on a and x in this form of rational polynomial will have the effect of flattening the central portion of the curve near $x = 0$. The

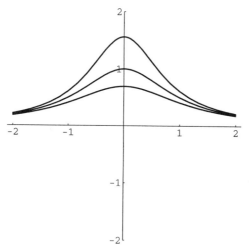

FIGURE 2.2.4. Graphs of $1/(a^2 + x^2)$ for $a = 0.8, 1.0, 1.2$.

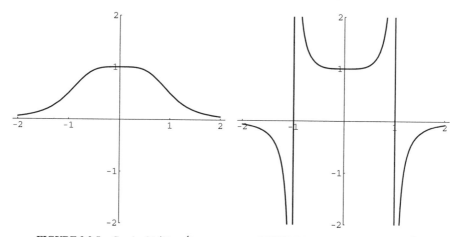

FIGURE 2.2.5. Graph of $1/(1 + x^4)$.   **FIGURE 2.2.6.** Graph of $1/(1 - x^4)$.

behavior for large x acts according to x^{-n} regardless of n, either from the positive or negative side of the x axis, depending on the sign of the x^n term in the denominator. There will always be just two poles for this form of the function when the sign in the denominator is negative, regardless of n. This is because the denominator in such a case can always be factored so that $-a$ and $+a$ are the only two real solutions to $h(x) = 0$.

Now let us take the above simple functions but make the numerator an even polynomial similar to the denominator. Consider this particular form:

$$y = c \frac{b^2 + x^2}{a^2 + x^2}$$

This function has no real poles and no real roots and is always positive in value if the

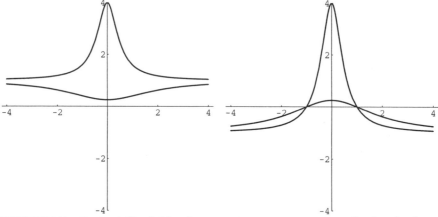

FIGURE 2.2.7.   Graphs of $(b^2 + x^2)/a^2 + x^2)$ for $b = 1, a = 0.5,\ 2.0$.

FIGURE 2.2.8.   Graphs of $(b^2 - x^2)/(a^2 + x^2)$ for $b = 1, a = 0.5,\ 2.0$.

sign of c is positive and always negative if not. What can we predict about this curve? The value at $x = 0$ is simply given by the constants as cb^2/a^2, and for large x it is given by c alone. Therefore, the height of the curve either rises or falls toward c as x goes away from the y axis, depending on the ratio b/a. In the limit of $b = 0$, the curve touches the x axis; in the limit of $a = 0$, the curve goes to infinity at $x = 0$. If $a = b$, the curve is simply a horizontal line at $y = c$. Figure 2.2.7 shows the realizations of this function for two values of b/a.

Let us introduce roots into the function at $-b$ and $+b$ by changing the sign of the x^2 term in the numerator; Figure 2.2.8 shows the resulting curves. Although the values of the curves at $x = 0$ are identical to those of the curves in Figure 2.2.7, the shapes are considerably different. These curves now always cross the x axis at the roots, $-b$ and $+b$, and thereafter remain negative, approaching $-c$ rather than $+c$ as in Figure 2.2.7.

By changing the sign of the x^2 term in the denominator to negative, we introduce poles at $-a$ and $+a$. The effect of these poles is shown in Figure 2.2.9. Note again that the vertical lines merely indicate the position of the poles and are not part of the curves themselves.

We give the last case of a ratio of two even binomials by having the signs of the x^2 term in both the numerator and the denominator be negative, thus introducing both poles and zeros into the function. Examples are plotted in Figure 2.2.10. Note the difference between the two examples; the shape depends on whether the pole lies inside the root (closer to the y axis) or outside of the root. Near $x = 0$, the behavior is governed by the value cb^2/a^2 as in all previous cases of this type; for large x, the value of the function again approaches c. By making the roots and the poles nearly coincident, the function will approach the constant c for all x except very near the roots and poles themselves. The plot in Figure 2.2.11 illustrates this.

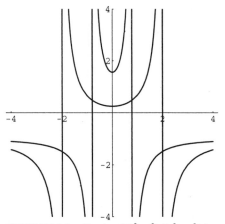

FIGURE 2.2.9. Graphs of $(b^2 + x^2)/(a^2 - x^2)$ for $b = 1, a = 0.8, 2.0$.

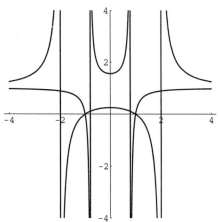

FIGURE 2.2.10. Graphs of $(b^2 - x^2)/(a^2 - x^2)$ for $b = 1, a = 0.8, 2.0$.

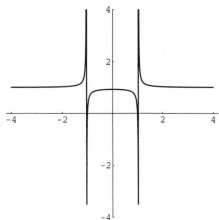

FIGURE 2.2.11. Graph of $(1 - x^2)/(1.04^2 - x^2)$.

FIGURE 2.2.12. Graph of $2(x + x^3)/(x + x^5)$.

An even rational polynomial can also be formed from the ratio of two odd polynomials. Consider the function

$$y = c \frac{x + x^3}{x + x^5}$$

Small x should behave as cx/x or simply as c; large x should behave as c/x^2. The graph of this function with $c = 2$ in Figure 2.2.12 confirms this.

2.2.4. Odd Rational Polynomials
The simplest form of an odd rational polynomial would be

$$y = \frac{b}{ax + x^3}$$

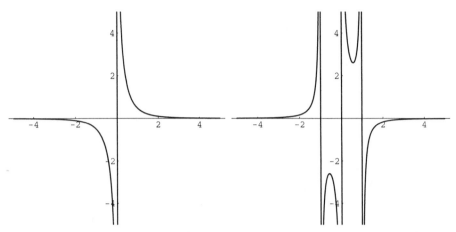

FIGURE 2.2.13. Graph of $1/(x+x^3)$.   **FIGURE 2.2.14.** Graph of $1/(x-x^3)$.

or

$$y = \frac{b}{ax - x^3}$$

Note that x can be factored out in both the denominators. The first function has only one pole at $x = 0$. The second has poles at $x = -a$ and $+a$ in addition to the one at $x = 0$. The graphs are radically different, as shown in Figures 2.2.13 and 2.2.14.

An odd rational polynomial will also result from a combination of even and odd polynomials: one in the numerator and the other in the denominator. Consider, for instance, the form

$$y = \frac{a + bx^3}{cx + x^3}$$

Let $a = 0$ so that the numerator is zero at $x = 0$. A plot of one particular instance of this function appears in Figure 2.2.15. The graph in this figure is antisymmetric, as predicted; but why didn't a pole appear at $x = 0$, a root of the denominator? Recall that the behavior of a polynomial at $x = 0$ is governed by the term with the smallest exponent; in the case of the denominator $x+x^3$, it is simply x. Thus the entire function behaves as bx^2/x or bx near $x = 0$. This is evident from the graph in Figure 2.2.15.

Let us take the reciprocal of this rational function to get

$$y = \frac{cx + x^3}{a + bx^2}$$

Again, letting $a = 0$, and using the same coefficients as for the curve above, we get the plot shown in Figure 2.2.16. Analysis of the dominant terms of both polynomials as "x approaches zero" explains the behavior of this graph near $x = 0$. In this vicinity, the numerator is approximated by x while the denominator is $5x^2$; thus the total effect is like $1/(5x)$, and there is a discontinuity at $x = 0$. For large x, however, the numerator behaves as x^3; and the total effect is then $x/5$.

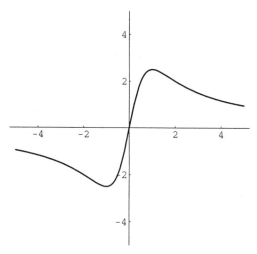

FIGURE 2.2.15. Graph of $5x^2/(x + x^3)$.

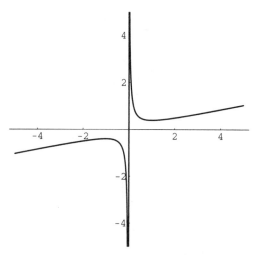

FIGURE 2.2.16. Graph of $(x + x^3)/5x^2$.

2.2.5. Factors of x^n

By itself, the factor x^n acts as either a zero of the function if in the numerator or as a pole of the function if in the denominator. If n is even, then it does not affect the symmetry of the function. If n is odd, it will change the symmetry, either from even to odd or odd to even. Let us see what these factors do to some of the rational polynomials treated thus far. First consider the form

$$y = \frac{b}{a^2 + x^2}$$

This is a symmetrical function. The graph in Figure 2.2.17 shows the effect of putting x in either the numerator (a zero) or the denominator (a pole) along with the original

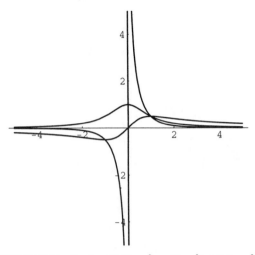

FIGURE 2.2.17. Graphs of $1/(1+x^2), x/(1+x^2), 1/[x(1+x^2)]$.

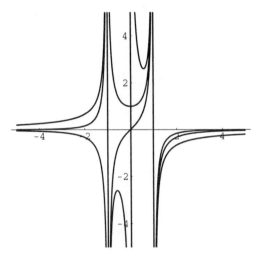

FIGURE 2.2.18. Graphs of $1/(1-x^2), x/(1-x^2), 1/[x(1-x^2)]$.

function. Note the antisymmetry imposed by the x factor in either case. Now take the same function, except change the sign of the x^2 term in the denominator to get

$$y = \frac{b}{a^2 - x^2}$$

The graph of this function and its variations when x is put first in the numerator and then in the denominator are shown in Figure 2.2.18 using $a = 1$. Note that the poles at $-a$ and $+a$ still influence the shape of the curves greatly.

Figures 2.2.19 and 2.2.20 illustrate what happens to these rational polynomials when the factor x^2 is applied to the numerator or denominator of the functions above, along with the graphs of the original functions.

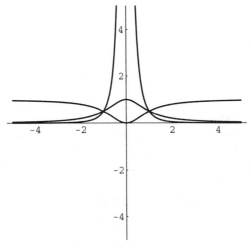

FIGURE 2.2.19. Graphs of $1/(1 + x^2)$, $x^2/(1 + x^2)$, $1/[x^2(1 + x^2)]$.

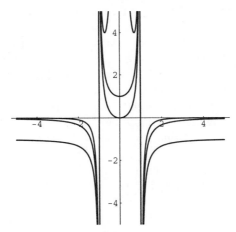

FIGURE 2.2.20. Graphs of $1/(1 - x^2)$, $x^2/(1 - x^2)$, $1/[x^2(1 - x^2)]$.

2.2.6. Behavior at Large and Small x

In studying the ordinary polynomial, we saw that the behavior at large x will depend solely on the term with the largest exponent. Likewise, for the rational polynomial, we need to consider only the terms with the largest exponents in predicting the behavior at large x.

Consider the form of the rational polynomial introduced at the beginning of Section 2.2.1. The behavior at large x reduces to the approximation $b_m x^m / a_n x^n$. This further reduces to $b_m x^{m-n} / an$. Therefore, for large x, we claim $f(x) \sim x^{m-n}$. For the special case that $m = n$, the function approaches a constant. If $m > n$, the function diverges to infinity. Lastly, if $m < n$, it goes to zero. This analysis is true regardless of the complexity of the polynomials in the numerator or denominator.

Similarly, the behavior at small x depends only on the terms with the smallest exponents m' and n'; again, this reduces to the general form $b_{m'} x^{m'-n'} / a_{n'}$. Therefore, $f(x) \sim x^{m'-n'}$ and the discussion relevant to large x can just as well be applied to small x.

2.3. SUMMARY

We have seen that algebraic analysis can give a fairly clear picture of the shape of a ordinary polynomial curve even before it is actually plotted. The position of roots and critical points and the large-x behavior together define the gross shape of the curve. We can vary the coefficients on the terms of different powers of x to produce predictable affects on the appearance of the curve, especially if the term has either the smallest or largest exponent.

Rational polynomial curves, having monomial terms in the denominator also, have considerably more variation in shape than ordinary polynomial curves. The key controllers of the shape of these curves are the zeros of the numerator and denominator. The latter are poles of the entire function forcing it to infinity unless they are cancelled by zeros of the numerator. Special cases of zeros and poles are factors of x^n that force the function to the origin if they are in the numerator or force it to infinity if in the denominator. The behavior of the function for large x was seen to depend on those terms with the largest exponents, both in the numerator and the denominator, while the behavior for small x similarly depends on those terms with the smallest exponents. If both have equal largest exponents, the curve approaches a constant for x large; this is true for small x in the case of equal smallest exponents.

Polynomials are quite flexible in their ability to produce varying curve shapes. For this reason, they are often used as fitting functions to give an approximation of data existing at regular or irregular increments of x or to smoothly approximate a series of control points which crudely define a desired shape. However, the extreme flexibility also tends to produce curves with more "shape" than required or desired, often showing large "overshoots" between the actual data or control points. These topics will be treated in detail in Chapter 11.

Chapter 3

RADICALS

3.1. FORM OF THE RADICAL

In the previous chapter, powers of constants or variables were introduced; for instance, the third power of x is represented by x^3. Normally, the power is an even positive integer; but it can be any real number. A simple function involving an arbitrary power is written as

$$y = x^p$$

The inverse operation of a power is a root. To express this generally, the number p is made the denominator of the exponent thus:

$$y = x^{1/p}$$

The notion of square root is universally recognized, and in this case $p = 2$. As for powers, the root can be an arbitrary real number.

The *radical* is a mathematical form indicating that a root is to be taken and is represented by the $\sqrt{}$ sign. To denote the exact root, the number p will precede the sign; for example, $\sqrt[3]{}$ denotes the third (or cube) root. The expression behind the radical sign is called the *radicand*. The following are all radical functions:

$$\sqrt[4]{(a + bx)}$$
$$\sqrt[4]{x^2}$$
$$\sqrt{(a/x)}$$
$$\sqrt{(3x + x^3 - 4x^5)}$$

By convention, if no number precedes the radical sign, it is understood to be the second (or square) root. In this chapter, we will only treat the square-root radical although the results can be easily, but not identically, extended to other roots. Any valid expression, algebraic or trigonometric, can be contained under the radical sign. This chapter will not treat any forms with trigonometric components, however. We will mainly discuss algebraic functions for which the radicand is a binomial raised to some power of x. These will be possibly multiplied by arbitrary radicals of x. Specifically, the general form to be studied in most of this chapter is

$$\sqrt{(a + bx^p)^n}\sqrt{x^m}$$

which can be also written as

$$(a + bx^p)^{n/2}x^{m/2}$$

Except for the last section, the variables m, n, and p are assumed to be integers throughout this chapter. We will further limit the integer value of p to either one

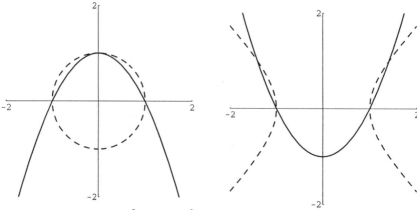

FIGURE 3.1.1.   Graphs of $1 - x^2$ and $\sqrt{(1 - x^2)}$.   **FIGURE 3.1.2.**   Graphs of $-1 + x^2$ and $\sqrt{(-1 + x^2)}$.

or two. Even with this limitation, the above form alone covers the large majority of radical functions commonly seen in mathematical engineering and physics, and it also covers a large suite of useful curve functions. With p equal to one or two and $m = 0$, the curves are, in fact, the familiar conic sections: parabolas, hyperbolas, and ellipses.

Note that the square-root radical can have both positive and negative roots. For instance, $\sqrt{4}$ is either -2 or $+2$ even though the positive square root is used in most applications. In the graphs of this section, we will see both the positive and negative branches. They are symmetric about the x axis; therefore, they are redundant in showing the shape of the curve. However, the dual branches often comprise the preferred presentation when one is interested in the shape of the curve. In fact, this symmetry provides useful curves for design.

How does the radical compare to the ordinary polynomial graphically? We will use the simple polynomial $1 - x^2$, which is a hyperbola, as an example. Note that it has roots at -1 and $+1$ because it factors into $(1 - x)(1 + x)$. Figure 3.1.1 shows the graph of $1 - x^2$ and the graph of $\sqrt{(1 - x^2)}$. The graph of the radical in this case is simply a circle of radius one.

The first property to note about this comparison is that the radical is confined to a limited domain of x, here from -1 to $+1$, while the domain of the polynomial itself is from minus infinity to plus infinity. When a polynomial, say $f(x)$, is put under the radical sign, only that part of the domain where $f(x) > 0$ can have the radical evaluated to a real number. We will call this region I of the radical. We can call the remainder of the domain, where $f(x) < 0$, region II. In this region, the radical function evaluates to an imaginary number and cannot be plotted. It is important to note that by putting a negative sign in front of $f(x)$, we swap region I for region II and produce a completely different graph for the radical. Using the example above, let $f(x) = -1 + x^2$ instead of $1 - x^2$; Figure 3.1.2 shows the new graphs. Note that the polynomial itself was simply reflected about the x axis. However, the graph of the radical changed from a circle to a pair of hyperbolas as region I became region II and vice-versa.

In the above general form of the function to be treated here, it is important to distinquish those cases based on whether m and n are even or odd integers.

1. *m even, n even*: This is the degenerate case because the radicals disappear. The function is equivalent to $(a + bx^p)^l x^k$, where $l = n/2$ and $k = m/2$. It is really an ordinary polynomial for which the discussion of the previous chapter applies. This case is not treated here.
2. *m even, n odd*: Here the x term is no longer a radical and can be rewritten as x^k where $k = m/2$. The remaining radical can be written as $(a + bx^p)^l \sqrt{(a + bx^p)}$ where $l = (n - 1)/2$. Note that l may be zero; in fact, this case is only treated for $l = 0$ in this chapter.
3. *m odd, n even*: Now the $(a + bx^p)$ radical simplifies to the nonradical $(a + bx^p)^l$ where $l = n/2$, and the $x^{m/2}$ factor decomposes to $x^k \sqrt{x}$ where $k = (m - 1)/2$. Note that k may be zero. This is the same as case (1) except for the $\sqrt{x}$ multiplier; in other words, it is essentially a polynomial with another factor of $\sqrt{x}$. Because $\sqrt{x}$ is not real for $x < 0$, the graph of the function is limited to $x > 0$. This case is not treated in this chapter.
4. *m odd, n odd*: Both the radicals remain. We can rewrite the expression as $(a + bx^p)^l x^k \sqrt{(a + bx^p)}\sqrt{x}$ where $l = (n - 1)/2$ and $k = (m - 1)/2$. Note that either k or l may be zero. Only the case for which both k and l are zero is treated in this chapter.

We will treat cases (2) and (4) separately. The graphs of each are distinctive enough that they can be considered as distinct functions. Within each case, we will treat $p = 1$ and $p = 2$ only. The important feature of these functions is that, due to the radical, they cannot exist over the entire x axis; only those regions of x where the radicands are positive will have a graph.

3.2. SIMPLE RADICALS

3.2.1. The Radical $\sqrt{(a + bx)}$

The graph of this function is equivalent to graphing $y^2 = a + bx$ or equivalent to reversing the dependency of the variables and graphing $x = (y^2 - a)/b$ instead. In this sense, we can clearly see that the function is a parabola. However, it is important to retain the conventional dependency of x and y variables and plot $y = f(x)$ which, in this case, involves the radical. It is clear that in order to have real values for the graph of the radical $\sqrt{(a + bx)}$ the binomial radicand must be greater than or equal to zero. This implies that the graph exists only within these limited domains of x for the four combinations of the signs of a and b:

1) $x > -a/b$ if $a > 0$, $b > 0$
2) $x < -a/b$ if $a < 0$, $b < 0$
3) $x > -a/b$ if $a < 0$, $b > 0$
4) $x < -a/b$ if $a > 0$, $b < 0$

The root of $a + bx$ lies at $-a/b$, thus the function equals zero there. Therefore, the graph lies wholly to one side or the other of the root. The examples using $|a| = 1$ and $|b| = 2$ in Figures 3.2.1 through 3.2.4 illustrate this.

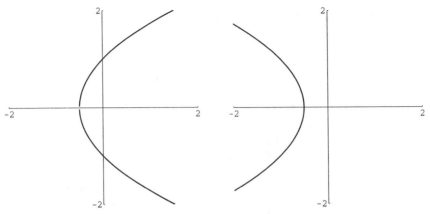

FIGURE 3.2.1. Graph of $\sqrt{(1 + 2x)}$.   **FIGURE 3.2.2.** Graph of $\sqrt{(-1 - 2x)}$.

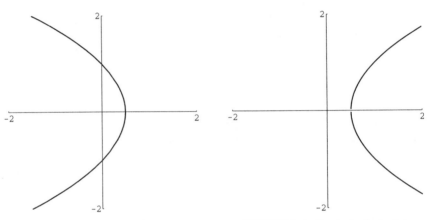

FIGURE 3.2.3. Graph of $\sqrt{(1 - 2x)}$.   **FIGURE 3.2.4.** Graph of $\sqrt{(-1 + 2x)}$.

Note that Cases (3) and (4) are the reflections about the y axis of Cases (1) and (2), respectively. This implies that really only two functions need to be considered insofar as the graphs are concerned.

3.2.2. The Radical $\sqrt{(a + bx^2)}$

For graphical purposes, this function is equivalent to $y^2 = a + bx^2$. We again determine the regions for which the radicand $(a + bx^2)$ is equal or greater than zero.

1) $-\infty < x < \infty$ if $a > 0, b > 0$
2) null   if $a < 0, b < 0$
3) $x < -\sqrt{(-a/b)}$ & $x > \sqrt{(-a/b)}$ if $a < 0, b > 0$
4) $-\sqrt{(-a/b)} < x < \sqrt{(-a/b)}$ if $a > 0, b < 0$

Cases 1, 3, and 4 are illustrated in Figures 3.2.5, 3.2.6, and 3.2.7, respectively, using $|a| = 1$ and $|b| = 2$.

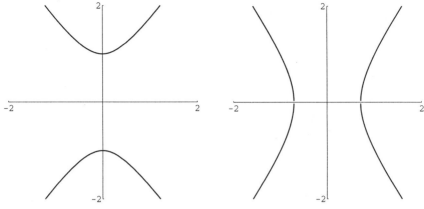

FIGURE 3.2.5. Graph of $\sqrt{(1 + 2x^2)}$.

FIGURE 3.2.6. Graph of $\sqrt{(-1 + 2x^2)}$.

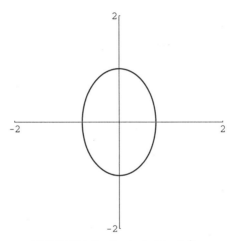

FIGURE 3.2.7. Graph of $\sqrt{(1 - 2x^2)}$.

Cases (1) and (3) are hyperbolas, but they differ in orientation. The one crosses the y axis at $\pm a$ while the other crosses the x axis at $\pm\sqrt{|a/b|}$. With a switch of constants and a rotation of 90 degrees, one case can be deformed into the other. Case (4) is the ellipse (or a circle when $b = -1$); it crosses the y axis at $\pm\sqrt{a}$ and the x axis at $\pm\sqrt{|a/b|}$. Therefore, the graph of Case (4) is tangent to the graphs of Cases (1) and (3) on the axes whenever the absolute values of the coefficients a and b are identical among the equations. Also, the graphs of Case (1) and Case (3) will approach one another for large values of both x and y if the absolute values of the coefficients are identical between the two equations. In a sense, Case (4) "inverts" to Cases (1) and (3) and does so by a change in the sign of the coefficient b.

3.2.3. The Radical $\sqrt{[x(a + bx)]}$
The graph of this equation is equivalent to $y^2 = ax + bx^2$. In order to have real values for the graph of this radical, the quantity $x(a+bx)$ must be greater than or equal to zero.

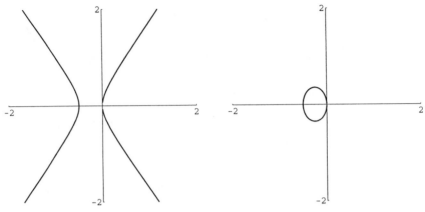

FIGURE 3.2.8. Graph of $\sqrt{[x(1+2x)]}$.   **FIGURE 3.2.9.** Graph of $\sqrt{[x(-1-2x)]}$.

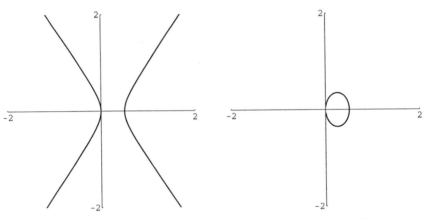

FIGURE 3.2.10. Graph of $\sqrt{[x(-1+2x)]}$.   **FIGURE 3.2.11.** Graph of $\sqrt{[x(1-2x)]}$.

Again, we identify the regions of x for which this is true under different combinations of the signs of a and b:

1) $x < -a/b$ & $x > 0$ if $a > 0$, $b > 0$
2) $-a/b < x < 0$ if $a < 0$, $b < 0$
3) $x < 0$ & $x > -a/b$ if $a < 0$, $b > 0$
4) $0 < x < -a/b$ if $a > 0$, $b < 0$

The graphs in Figures 3.2.8 through 3.2.11 illustrate these four cases in order.

Note that the graphs of Cases (3) and (4) are the reflections about the y axis of Cases (1) and (2), respectively. This can be verified by substituting $-x$ for x in the expressions for (1) and (2). Thus, graphically, we have really just two forms to consider.

Recall that for polynomial functions the behavior of the graph is controlled at small x by the x term with the smallest exponent and at large x by the x term with the largest exponent. The radical function behaves the same, except that the y values are now taken to some power n.

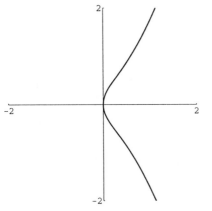

FIGURE 3.2.12. Graph of $\sqrt{[x(1 + 2x^2)]}$.

FIGURE 3.2.13. Graph of $\sqrt{[x(-1 - 2x^2)]}$.

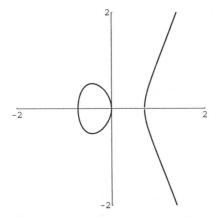

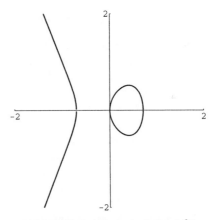

FIGURE 3.2.14. Graph of $\sqrt{[x(-1 + 2x^2)]}$.

FIGURE 3.2.15. Graph of $\sqrt{[x(1 - 2x^2)]}$.

3.2.4. The Radical $\sqrt{[x(a + bx^2)]}$

The graph of this radical is equivalent to plotting $y^2 = ax + bx^3$. Again, we define the regions of x for which this radical is positive for the different combinations of signs of a and b:

1) $x > 0$ if $a > 0, b > 0$
2) $x < 0$ if $a < 0, b < 0$
3) $-\sqrt{(-a/b)} < x < 0$ & $x > \sqrt{(-a/b)}$ if $a > 0, b < 0$
4) $x < -\sqrt{(-a/b)}$ & $0 < x < \sqrt{(-a/b)}$ if $a < 0, b > 0$

The graphs in Figures 3.2.12 through 3.2.15 illustrate these four cases in order.

Note that Case (2) is the reflection of Case (1) about the y axis and that Case (4) is the reflection of Case (3). These can be verified by substituting $-x$ for x in the form of the expressions for Cases (1) and (2). Thus, symmetry implies that there are, graphically, just two cases to consider.

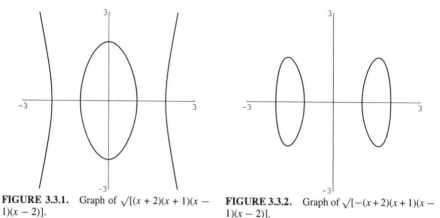

FIGURE 3.3.1. Graph of $\sqrt{[(x + 2)(x + 1)(x - 1)(x - 2)]}$.

FIGURE 3.3.2. Graph of $\sqrt{[-(x + 2)(x + 1)(x - 1)(x - 2)]}$.

3.3. ROOTS OF THE RADICAND

As for the polynomials, one can study the behavior of the radical function simply by looking at the roots of the radicand. First, one must recognize that for the radicand none of the roots can be double roots (or any other higher even multiplicity) or else the factors formed by the roots will be of even power and can consequently be brought outside of the radical sign. Second, if the roots occur only as even-odd pairs of real numbers, such as $(-1, +1)$, the graph of the radicand itself will necessarily be symmetric about the y axis. Consequently, the graph of the radical expression will be symmetric about the y axis.

We can further classify the graph of the radicals with real roots according to whether the total number of roots is even or odd. If the number is even, there is a further distinction between graphs for which the function of the roots' product is positive or negative.

Beginning with an even number of roots, we recall the graphs of Section 3.2.2. Depending on how the roots are defined, we either produce an ellipse or a hyperbola. The radicand for the ellipse has positive a and negative b (for example, $1 - x^2$), and the radicand for the hyperbola is negative a and positive b (for example, $x^2 - 1$). Both radicands have the same roots at -1 and $+1$, but the graphs are quite different. The difference lies in the sign of one of the factors formed based on the roots. For the root at $+1$, we can use the factor $x - 1$ or $1 - x$. Depending on which of these is multiplied with the factor due to the other root, $x + 1$, the hyperbola or ellipse is generated. In general, we can make either closed curves containing the roots or curves which go away from the roots simply by changing the signs in a factor.

We might suspect that this works for higher-degree radicands. Consider these roots: $-2, -1, +1, +2$. If the factors are formed normally, we get $(x + 2)(x + 1)(x - 1)(x - 2)$ for the radicand; the graph appears in Figure 3.3.1 when the square root is taken. The outer curves go away from the roots; we expect this because the large x behavior should be given by the term in the radicand with the largest power of x. Now, change both the signs in the factor for any one of the roots. This is equivalent to putting a negative sign in front of the entire expression. The resulting graph is shown in Figure 3.3.2; note that the pieces of this graph fit between those of Figure 3.3.1.

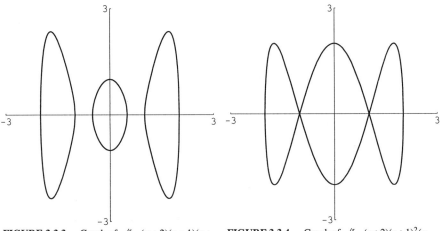

FIGURE 3.3.3. Graph of $\sqrt{[-(x+2)(x+1)(x+1/2)(x-1/2)(x-1)(x-2)]}$.

FIGURE 3.3.4. Graph of $\sqrt{[-(x+2)(x+1)^2(x-1)^2(x-2)]}$.

Adding two roots at $x = 1/2$ and $-1/2$ produces the graph shown in Figure 3.3.3. Using two roots at $x = 1$ and $x = -1$ instead of at $1/2$ and $-1/2$ produces the graph in Figure 3.3.4. Note that the multiple roots cause the ellipse-like closed curves to become loops of one continuous curve.

The pattern that has emerged is that for radicands with an even number of real roots the graph of the radical takes one of two basic forms:

1. If the n roots are used in the conventional manner as factors $(x-r)$ and the radicand is taken as the product, the graph has two hyperbolic-like branches going away from the two outlying roots and $(n-2)/2$ ellipse-like closed curves each passing through pairs of adjacent remaining roots.

2. If the n roots are used in the conventional manner as factors $(x-r)$ and the radicand is taken as the negative of the product (equivalent to negating any one of the factors), the graph has $n/2$ ellipse-like closed curves each passing through pairs of adjacent roots.

Now let us examine radicands with an odd number of roots. Again, as with an even number of roots, we can form a new graph if the negative of the product of the root factors is taken. Consider the graph of the function formed from the roots -2, $+1$, and $+2$ (Figure 3.3.5). If we negate one of the factors or equivalently negate the entire radicand, the graph in Figure 3.3.6 is the result. Adding two more roots at $x = 0$ and $x = -1$ produces the graph in Figure 3.3.7.

The pattern that has emerged for radicands with an odd number of real roots is that:

1. If the n roots are used in the conventional manner as factors $(x-r)$ and the radicand is taken as the product, the graph has a hyperbolic-like branch going away from the right-most root and $(n-1)/2$ ellipse-like closed curves each passing through pairs of adjacent remaining roots.

2. If the n roots are used in the conventional manner as factors $(x-r)$ and the radicand is taken to be the negative of the product (equivalent to negating any one of the

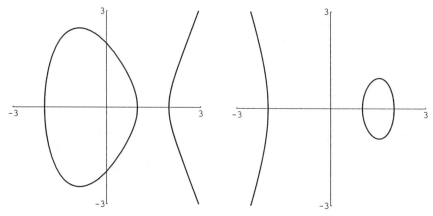

FIGURE 3.3.5. Graph of $\sqrt{[(x+2)(x-1)(x-2)]}$.   **FIGURE 3.3.6.** Graph of $\sqrt{[-(x+2)(x-1)(x-2)]}$.

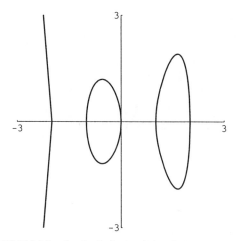

FIGURE 3.3.7. Graph of $\sqrt{[-(x+2)(x+1)x(x-1)(x-2)]}$.

factors), the graph has a hyperbolic-like branch going away from the left-most root and $(n-1)/2$ ellipse-like closed curves each passing through pairs of adjacent remaining roots.

As for polynomials, we can see that the shape of the curve can be controlled by placement of the roots of the radicand. In addition, we can generate curves with different features by using multiple roots.

While the polynomial expressions always produced continuous curves, the radicals do not necessarily do so. This is due simply to the fact that the radicand becomes negative over some of the domain of x, Consequently, the square-root radical cannot be evaluated to a real number. This would be true for any order of radical where the root is an even power. (Recall that it is assumed to be two throughout this chapter.) However, if the root were an odd power, the root of a negative radicand will also be

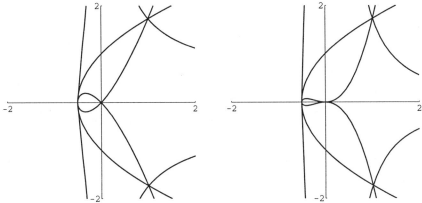

FIGURE 3.4.1. Graphs of $x^n \sqrt{(1 + 2x)}$ with $n = -1, 0, 1$.

FIGURE 3.4.2. Graphs of $x^n \sqrt{(1 + 2x)}$ with $n = -2, 0, 2$.

evaluated as real. Thus, for instance, we would evaluate cube-root expressions such as $\sqrt[3]{(1 + x)}$ over the entire x axis as a real number; it would then be plotted as a continuous curve.

3.4. THE FACTOR x^n

Let us consider the effect of multiplying the radicals studied above with x^n where n is an integer. From previous sections, we have seen that the effect of x^n is to create a zero, or root, of the function at $x = 0$ or, if n is negative, a pole of the function at $x = 0$. If n is even, the sign of the function is unchanged for both $x < 0$ and $x > 0$; thus any symmetry is preserved. But, if n is odd, the function will change its sign from greater than zero to less than zero, or vice-versa, at $x = 0$.

In considering the effect of x^n, those radicals which have graphs entirely on one side or the other of the y axis and do not touch it will not be of interest because they are not strongly affected by the root or pole created with this factor. Let us look at some radicals that are strongly affected though, beginning with $\sqrt{(a + bx)}$, which is the parabola. Figure 3.4.1 shows the original parabola and also the result of multiplying and dividing it by x. Note that x in the numerator (root) causes the parabola to form a loop and that x in the denominator (pole) causes the parabola to stretch out to infinity at $x = 0$. Figure 3.4.2 shows the original parabola and also the effect of both multiplying and dividing it by x^2.

Next, consider the hyperbola which was seen above in the form $\sqrt{(a + bx^2)}$ with a and b both positive. Figures 3.4.3 and 3.4.4 show this hyperbola and the effect of multiplying and dividing it by x and x^2, respectively.

Now consider the ellipse which appeared in the form $\sqrt{(a + bx^2)}$ with $a > 0$ and $b < 0$. Figures 3.4.5 and 3.4.6 show the original ellipse and the modifications produced by multiplying or dividing by x and x^2, respectively.

Next, consider the radical $\sqrt{[x(a + bx)]}$ which already has a root at $x = 0$. The first graphical form appears when either $a > 0$ and $b > 0$ or when $a < 0$ and $b > 0$ (the latter is the reflection of the former about the y axis). Figures 3.4.7 and 3.4.8 show

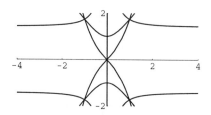

FIGURE 3.4.3. Graphs of $x^n \sqrt{(1+2x^2)}$ with $n = -1, 0, 1$.

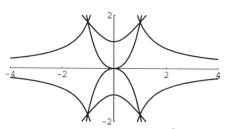

FIGURE 3.4.4. Graphs of $x^n \sqrt{(1+2x^2)}$ with $n = -2, 0, 2$.

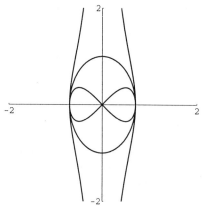

FIGURE 3.4.5. Graphs of $x^n \sqrt{(1-2x^2)}$ with $n = -1, 0, 1$.

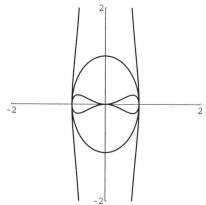

FIGURE 3.4.6. Graphs of $x^n \sqrt{(1-2x^2)}$ with $n = -2, 0, 2$.

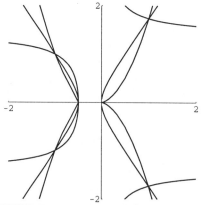

FIGURE 3.4.7. Graphs of $x^n \sqrt{[x(1+2x)]}$ with $n = -1, 0, 1$.

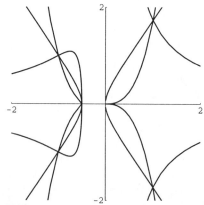

FIGURE 3.4.8. Graphs of $x^n \sqrt{[x(1+2x)]}$ with $n = -2, 0, 2$.

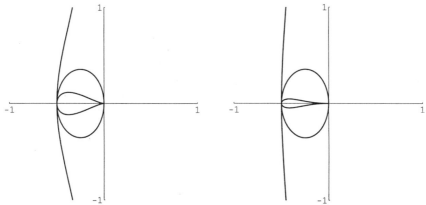

FIGURE 3.4.9. Graphs of $x^n \sqrt{[x(-1-2x)]}$ with $n = -1, 0, 1$.

FIGURE 3.4.10. Graphs of $x^n \sqrt{[x(-1-2x)]}$ with $n = -2, 0, 2$.

the original function and modifications produced by multiplying or dividing by x and x^2, respectively.

Consider the same radical $\sqrt{[x(a+bx)]}$ but now when either $a < 0$ and $b < 0$ or $a > 0$ and $b < 0$ (the latter is just the reflection of the former about the y axis). Figures 3.4.9 and 3.4.10 illustrate the effect of multiplying and dividing by x and x^2, respectively.

3.5. A VARIATION ON CONIC SECTIONS

The equations thus far in this chapter have been fairly restrictive in regard to the exact numbers used in exponents and for roots in the general radical form. This section focuses on just one departure which allows some special curves of interest to be generated. The equation

$$x^2/a^2 + y^2/b^2 = 1$$

gives the standard ellipse with axes of length $2a$ and $2b$. This can be rewritten as the equation

$$y = [1 - c^2 x^2]^{1/2}$$

where $c = b/a$. Note that the denominator of the root is equal to the exponent of x. If we generalize this number to q, with q arbitrary, the equation is

$$y = [1 - |cx|^q]^{1/q}$$

The reason for taking the absolute value of cx is to preserve the symmetry about the y axis in all cases. For $q = 2$, we have the normal ellipse; for $q > 2$, we have a *hyperellipse*. The graph of this function is shown in Figure 3.5.1 for $q = 2, 3, 4, 5$ and $c = 1/2$. As q becomes large, the hyperellipse will approach the shape of a rectangle.

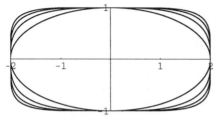

FIGURE 3.5.1.   Ellipse ($q = 2$) and hyperellipses for $q = 3, 4, 5$.

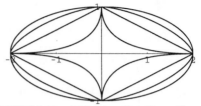

FIGURE 3.5.2.   Ellipse ($q = 2$) and hypoellipses for $q = 3/2, 1, 1/2$.

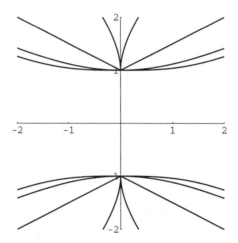

FIGURE 3.5.3.   Hyperbola ($q = 2$) and its variations for $q = 3, 1, 1/2$.

For the hyperellipse, the value of q could be any real number greater than two, not necessarily an integer, although the examples used integer values. If we allow q to be less than two, the figure is called a *hypoellipse*. Examples are shown in Figure 3.5.2 for $q = 3/2$, 1, and 1/2 along with the normal ellipse for $q = 2$. For $q = 1$, the figure is a diamond composed of simply straight lines.

A similar deformation can be made to the hyperbola which is given by

$$x^2/a^2 - y^2/b^2 = 1$$

In this case, the equation to be graphed is

$$y = [1 + |cx|^q]^{1/q}$$

The resulting curves are not as interesting or useful as those related to the ellipse; they are shown in Figure 3.5.3 for $c = 1/2$ again and $q = 1/2$, 1, 2, 3. The case $q = 2$ is the normal hyperbola, and the case $q = 1$ gives simple straight lines like the same case for the hypoellipse.

3.6. SUMMARY

Most of the radicals introduced in this chapter, when plotted, were either the familiar conic sections (parabola, hyperbola, or ellipse) or variations of them. In general, the graphs of these radicals are limited to certain regions of the x domain due to the fact that the radicand must remain positive in order for a square-root to evaluate to a real number. These regions are identified with a simple analysis of the function and are always bracketed by any roots of the radicand. When these radicals are multiplied or divided by powers of x, a root or pole, respectively, is introduced at $x = 0$; this leads to predictable modifications of the graph of the radicals. If the radicands can be factored into binomials of the form $(x - r)$, then the graphs are fairly predictable based on the roots and whether there are an even or odd number of them.

Chapter 4

EXPONENTIAL FUNCTIONS

4.1. INTRODUCTION TO THE EXPONENTIAL FUNCTION

4.1.1. Elementary Exponential Function

The elementary *exponential function* is $y = e^x$. This function is widely used in pure and applied mathematics from basic number theory to advanced calculus. The number e is the limit of

$$(1 + 1/n)^n$$

as n becomes large. It is an irrational number, and its approximate value is 2.71828. It is also the base of the natural logarithms; we express them as $y = \log_e x$ or $y = \ln(x)$. Figure 4.1.1 shows a plot of the exponential function.

We can appreciate, from the right side of this graph, the meaning of "exponential growth." On the other side of the y axis, we see "exponential decay." The growth and decay can be switched about the axis by merely changing the sign of x. Note that the limits of the exponential curve are zero at negative x infinity and infinity at positive x infinity. The value at $x = 0$ is unity. This last statement is true regardless of whether a constant other than unity is used to multiply x because $e^0 = 1$.

Perhaps the most important property of the exponential function, from an applied mathematics viewpoint, is that all derivatives of e^x are equal to e^x. This means the slope of the function equals the function itself, the rate of change of the slope equals the slope itself, and so on. Thus, for large x, the path to infinity becomes increasingly steep. On the other side of the y axis, the asymptotic value of zero is approached slowly as the function becomes nearly flat.

In using the exponential function, we usually include some constant as a factor of x:

$$y = e^{ax}$$

Values of a greater than unity increase the slope of the curve everywhere while values less than unity decrease it everywhere. In general, the derivative of the exponential e^{ax} is simply ae^{ax}. Each derivative is therefore a times the previous derivative so that the function is no longer invariant with differentiation whenever $a \neq 1$.

4.1.2. Exponential Functions with More Complicated Exponents

Two variations of the exponential function are of interest because of their symmetry. These functions are

$$y = \exp(a|x|)$$
$$y = \exp(ax^2)$$

We have introduced the notation "$\exp[f(x)]$" meaning e with the exponent $f(x)$. The first of the two functions is plotted in Figures 4.1.2 and 4.1.3 with $a = +1$ and -1, respectively.

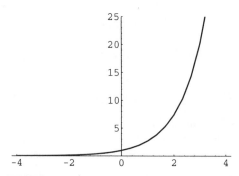

FIGURE 4.1.1. Graph of the exponential function e^x.

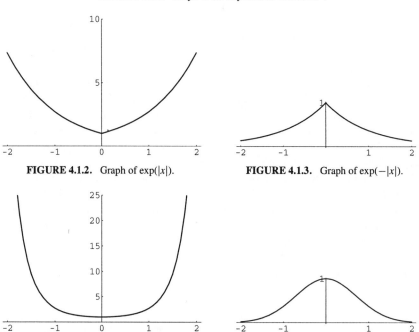

FIGURE 4.1.2. Graph of $\exp(|x|)$.

FIGURE 4.1.3. Graph of $\exp(-|x|)$.

FIGURE 4.1.4. Graph of $\exp(x^2)$.

FIGURE 4.1.5. Graph of $\exp(-x^2)$.

These curves are actually the left and right halves of the original function e^{ax}, each folding symmetrically about the y axis; therefore, they represent nothing new. The second of the two functions is plotted in Figures 4.1.4 and 4.1.5, again with $a = +1$ and -1, respectively.

The function $\exp(ax^2)$, when used with negative a, is commonly called the *normal curve*, the *bell curve*, or the *Gaussian curve*. The width of the curve can easily be altered by putting a constant other than unity before the x^2 argument as shown by the graphs in Figure 4.1.6 for values of $a = 1, 2, 3$.

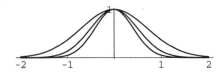

FIGURE 4.1.6. Graphs of $\exp(-ax^2)$ for $a = 1, 2, 3$.

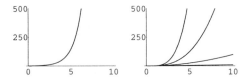

FIGURE 4.1.7. Graphs of e^x and x^n for $n = 1, 2, 3, 4$.

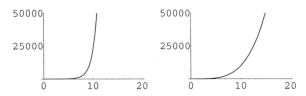

FIGURE 4.1.8. Graphs of e^x and x^4.

4.1.3. Comparison with the Monomial x^n

It is important to present the mathematical relation between the exponential function and the monomials x^n. We can write the former as an infinite sum of the latter, thus:

$$e^x = \sum_{n=1}^{\infty} \frac{x^n}{n!} \quad (-\infty < x < \infty)$$

Recall that in our discussion of polynomials it was shown that for large x the value of the polynomial is dominated by that term with the largest exponent. Similarly, the value of e^x, at large x, should be dominated by x^∞. For positive x, it is straightforward to claim that the value of e^x exceeds that of x^n for $n < \infty$ at large x. However, if x is negative, we see from the above series that the terms will alternate in sign. For this special series, it can be shown that the sum of infinite terms for negative x will be zero. These limits for e^x in the positive and negative domains of x are clear from the plot of the exponential function in Figure 4.1.1.

The behavior of the exponential function must, therefore, be dominant at large positive x when it is a factor along with polynomials in an expression. We can show this by considering the general multiplicative monomial x^n. Consider the function $x^n \exp(x)$ for increasing values of n. Figure 4.1.7 shows the graph of the two factors of this function (x^n for $n = 1$ to 4). Here, we see that x^4 exceeds e^x at $x = 4$. However, we can always expand the region of the plot to show that e^x dominates at large x. In Figure 4.1.8, we have increased the y range to 50,000. Now, e^x is larger than x^4 beginning at approximately $x = 10$.

For negative arguments, we can equivalently consider the expression $x^n \exp(-x)$ for positive values of x. Will the graph of this function be dominated by $\exp(-x)$

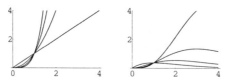

FIGURE 4.1.9. Graphs of e^x and $x^n e^{-x}$ for $n = 1, 2, 3, 4$ on the domain [0, 4].

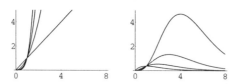

FIGURE 4.1.10. Graphs of e^x and $x^n e^{-x}$ for $n = 1, 2, 3, 4$ on the domain [0, 8].

FIGURE 4.2.1. Graphs of first three derivatives of Gaussian function.

and tend towards zero for large x? Figure 4.1.9 shows the graph of x^n alone and x^n multiplied by $\exp(-x)$ ($n = 1$ to 4). In Figure 4.1.9, it is clear that the curve is tending to zero for $n = 1, 2, 3$ but not necessarily for $n = 4$. The remedy is to merely extend the domain of x upward as in Figure 4.1.10. We will find that for sufficiently large x, no matter what exponent x^n has, the negative exponential factor always drives the curve to $y = 0$ at large x.

4.2. DERIVATIVES OF THE GAUSSIAN CURVE

The derivatives of the Gaussian curve form an interesting family of curves. The first three derivatives are given here for $a = 1$:

$$-2x \exp(-x^2)$$
$$(-2 + 4x^2) \exp(-x^2)$$
$$(12x - 8x^3) \exp(-x^2)$$

The graphs of the first three derivatives, as given in the equations above, appear in Figure 4.2.1 with y scaling factors of 4, 2, and 1, respectively.

The higher derivatives continue to expand the number of terms in the polynomial factor: each odd derivative has a polynomial of all odd terms and each even derivative has one of all even terms. This family of curves then has alternating odd and even functions. There is a progression in the number of zero crossings with derivative number. This number is equal to the order of the derivative. Thus, we can roughly predict the form of the curve for an arbitrary order of the derivative.

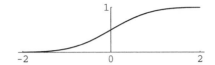

FIGURE 4.3.1. Graph of the integral of the Gaussian curve.

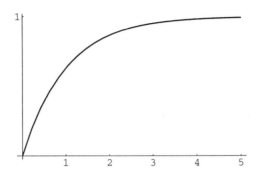

FIGURE 4.4.1. Graph of exponential ramp $1 - e^{ax}$ with $a = -1$.

4.3. INTEGRALS OF THE GAUSSIAN CURVE

The integral of the Gaussian function, with a constant factor of $\pi^{-1/2}$, is defined by

$$\frac{1}{\sqrt{\pi}} \int_{-\infty}^{x} e^{-t^2} dt$$

This function is plotted in Figure 4.3.1. We can easily alter the shape of this function by using a constant multiplier on x in the definition. Note that it is asymptotic to $y = 0$ and $y = 1$ for large negative x and large positive x, respectively. This curve has a pleasing "S" shape that may be useful in many design applications. The integral of this function, or the second integral of the Gaussian curve, will be zero at negative infinity along the x axis and will diverge to infinity at large positive x.

4.4. OTHER USEFUL FUNCTIONS INVOLVING THE EXPONENTIAL

A few functions involving the exponential in more complicated ways are worthy of examination. First, consider this function:

$$f(x) = 1 - e^{ax}$$

When the coefficient a is negative, this function is called the *exponential ramp*. The example in Figure 4.4.1 should explain why it is called such. When used as a multiplicative factor in the positive x domain only, this function provides a nice means to "ramp up" another curve from the origin. The rate of the ramping can be controlled by the coefficient a.

Next, consider the following function:

$$f(x) = 1/(1 + e^{ax})$$

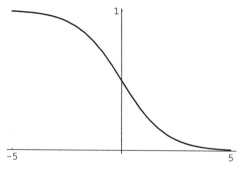

FIGURE 4.4.2. Sigmoidal curve with $a = 1$.

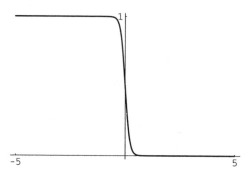

FIGURE 4.4.3. Sigmoidal curve with $a = 10$.

By applying what we already know about the exponential function, we should be able to describe the graph of this particular function. Because $e^0 = 1$, this curve will cross the y axis at a value of $y = 1/2$. Because the limit of the exponential function is infinity at large positive x, $f(x)$ approaches zero there. Because the limit is zero at large negative x, $f(x)$ approaches unity there. The curve is known as the *sigmoidal* curve F and is shown in Figure 4.4.2 for $a = 1$. The slope of the step down from unity to zero can be modified by placing a constant multiplier on x. Extremely large values of a will produce nearly a vertical step as seen in Figure 4.4.3.

Another interesting set of exponential curves is produced from the function

$$f(x) = ae^{-bx} - ce^{-dx}$$

when it is evaluated over the range of positive x with all the constants having positive values. An example is shown in Figure 4.4.4 for $a = 1$, $b = 1$, $c = 1$, and $d = 3$. Note that both exponential terms are of the exponential decay type. The term with the larger constant will diminish more rapidly, and the function becomes nearly equal to the other term alone. Because both terms equal unity at $x = 0$, the difference must equal zero as shown. In between, the function rises to a peak and then starts to decay. We can determine the position of the peak by setting the derivative of the function equal to zero, thus:

$$-abe^{-bx} + cde^{-dx} = 0$$

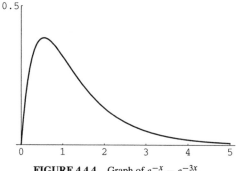

FIGURE 4.4.4. Graph of $e^{-x} - e^{-3x}$.

and we solve for x to get

$$x_{peak} = [\ln(cd) - \ln(ab)]/(d - b)$$

For the above plot, $x_{peak} \approx 0.55$. The function for which we have plotted one example in Figure 4.4.4 is useful in making arbitrary "pulse-like" curves that rise and then decay with a desired shape.

4.5. SUMMARY

In this chapter, we have seen that exponential curves form a distinct family. Either they have a shape that decays to zero with large x or a shape that accelerates rapidly to infinity at large x depending on the sign of the exponent. Due to this property, we can use exponentials to "ramp up" or "ramp down" other types of functions. When combined with polynomial factors, the exponential always dominates. A useful set of the exponential functions are the Gaussian and related curves which decay on both sides of the y axis to give either symmetric or antisymmetric wavelet-like functions that are appreciably nonzero only over a small domain of x. More interesting, wavelet-like functions can be constructed with the sum of two exponential functions.

Chapter 5

PERIODIC FUNCTIONS ON A LINE

5.1. BASICS OF THE PERIODIC FUNCTIONS

This chapter will present the graph of the periodic sine (or cosine) function and several of its variants. The origin of the periodic function rests in trigonometry where the sine and cosine functions are defined. The familiar diagram in Figure 5.1.1 should help you to recall these definitions. Consider, if the vector r starts parallel to the x axis ($\theta = 0$) and then rotates through a full circle back to the x axis ($\theta = 2\pi$). During this revolution, the values of sine and cosine vary continuously between the limits of -1 and $+1$; the graphs of the functions appear in Figure 5.1.2.

Here, we have merely stretched the angle θ out onto a horizontal axis. It should be clear that the entire behavior of these two functions is illustrated by this plot from 0 to 2π. If the vector r were to continue rotating past the horizontal axis, the functions will repeat exactly for each 2π revolution. Because they repeat endlessly, we call them *periodic* functions. A periodic function is defined mathematically as

$$f(\theta + L) = f(\theta)$$

where L is the *fundamental period*, which is 2π for the elementary sine and cosine functions. Although the trigonometric functions are not the only periodic functions, they are by far the dominant ones used in practice.

The sine and cosine functions are actually special cases of the more general periodic function

$$\sin(a\theta + b)$$

which, by a fundamental trigonometric identity, can be expanded into

$$\sin(a\theta)\cos(b) + \cos(a\theta)\sin(b)$$

Because b is a constant, we can set $c = \cos(b)$ and $d = \sin(b)$ to obtain

$$c\sin(a\theta) + d\cos(a\theta)$$

Note that, again by a fundamental trigonometric identity, $c^2 + d^2 = 1$. The special cases which are the sine and cosine function arise when

$$c = 1, d = 0 \rightarrow \sin(a\theta)$$
$$c = 0, d = 1 \rightarrow \cos(a\theta)$$

Figure 5.1.3 shows $\sin(a\theta + b)$ for $a = 1$ and $b = 0$ to $3\pi/4$ with an increment of $\pi/4$. Note that the first and third of these functions are the ordinary sine and cosine functions, respectively. Clearly, we could generate a continuum of functions by letting b vary continuously between 0 and 2π.

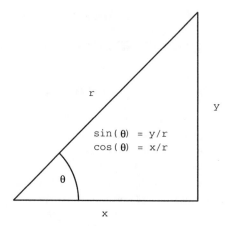

FIGURE 5.1.1. Geometrical definition of sine and cosine functions.

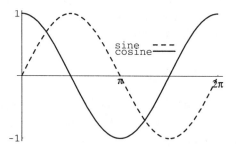

FIGURE 5.1.2. Graphs of sine and cosine functions.

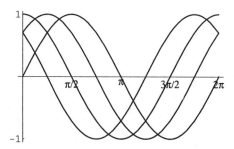

FIGURE 5.1.3. Graphs of $\sin(2\pi\theta + b)$ for $b = 0, \pi/4, \pi/2, 3\pi/4$.

In Chapter 2, we defined even and odd functions. According to this definition, the sine function is odd and the cosine function is even. An additional property of periodic functions is referred to as even or odd harmonic. If L is the period of the function and $f(\theta + L/2) = -f(\theta)$, then the function is *odd harmonic*. If $f(\theta + L/2) = +f(\theta)$, then it is *even harmonic*. In the case of an even harmonic, the function is actually periodic with fundamental period $L/2$. It is important to recognize the following relations in the multiplication of even and odd functions:

$$\text{even} * \text{even} = \text{even}$$
$$\text{odd} * \text{odd} = \text{even}$$
$$\text{even} * \text{odd} = \text{odd}$$
$$\text{odd} * \text{even} = \text{odd}$$

The fundamental period is unchanged in all cases.

In this chapter, we will study both ordinary and parametric forms for periodic functions. By an "ordinary" form, we mean a form

$$y = f(x)$$

where f is periodic. Only *single-valued functions* f are allowed; that is, for each x value, only one y value exists. By a "parametric" form, we mean a form

$$x = f(t); \; y = g(t)$$

where both f and g are periodic. The parametric forms can lead to some interesting and intricate graphs due to the added flexibility of this form; they can produce *multivalued functions* where more than one y value can exist at certain values of x.

5.2. SIMPLE PERIODIC FUNCTIONS

5.2.1. Compound Harmonics

A *compound harmonic* is expressed as one of the following:

1) $y = \sin[a\pi \sin(2\pi x)]$
2) $y = \sin[a\pi \cos(2\pi x)]$
3) $y = \cos[a\pi \cos(2\pi x)]$
4) $y = \cos[a\pi \sin(2\pi x)]$

The graphs of (1) and (2) are identical except for a phase shift of $\pi/2$; this is also the case for (3) and (4). Consider the graph of (1) for $a = 1/4, 1/2, 1$, and 2 as shown in Figures 5.2.1 to 5.2.4, respectively. The fundamental period of one is unchanged while the curves exhibit increasing complexity as the constant a increases. We know from calculus that for small argument p

$$\sin(p) \sim p$$

This explains why the compound harmonic appears to approach $2a\pi^2 x$, a line with slope $2a\pi^2$, as x becomes small.

Now let us look at the graph of the compound harmonic in Case 3 above for again $a = 1/4, 1/2, 1$, and 2 as shown in Figures 5.2.5 to 5.2.8. The behavior of the compound cosine harmonic is very different than that of the compound sine harmonic. However, it again exhibits more complexity as the constant a is increased. Also note

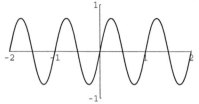

FIGURE 5.2.1. Graph of $\sin[\pi/4\sin(2\pi x)]$.

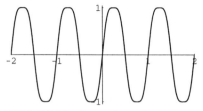

FIGURE 5.2.2. Graph of $sin[\pi/2\sin(2\pi x)]$.

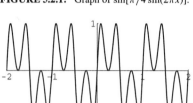

FIGURE 5.2.3. Graph of $\sin[\pi\sin(2\pi x)]$.

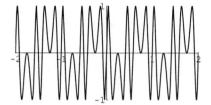

FIGURE 5.2.4. Graph of $\sin[2\pi\sin(2\pi x)]$.

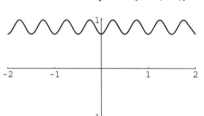

FIGURE 5.2.5. Graph of $\cos[\pi/4\cos(2\pi x)]$.

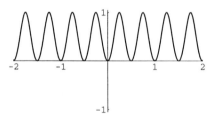

FIGURE 5.2.6. Graph of $\cos[\pi/2\cos(2\pi x)]$.

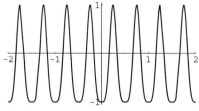

FIGURE 5.2.7. Graph of $\cos[\pi\cos(2\pi x)]$.

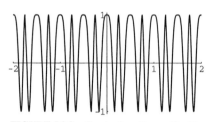

FIGURE 5.2.8. Graph of $\cos[2\pi\cos(2\pi x)]$.

that the fundamental period has changed to 1/2, compared to 1 for the compound sine wave. From calculus, we know that for small argument p

$$\cos(p) \sim 1$$

Thus, as $a \rightarrow 0$, the argument of the compound cosine harmonic goes toward the value of zero regardless of x. This explains the diminishing of the oscillations and the approach toward unity as the constant a is decreased.

The compound harmonics will reveal many interesting curves as the constant a is varied. The fundamental period always is equal to the constant in the argument of the inner harmonic divided by 2π for the compound sine and 4π for the compound cosine, regardless of the value of a.

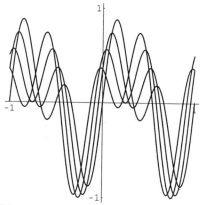

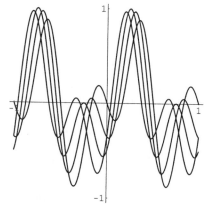

FIGURE 5.2.9. Graph of sine wave plus second harmonic ($b = 0$, $\pi/4$, $\pi/2$, $3\pi/4$).

FIGURE 5.2.10. Graph of sine wave plus second harmonic ($b = \pi$, $5\pi/4$, $3\pi/2$, $7\pi/4$).

5.2.2. Sums of Harmonics

If an elementary sine function has a period of L, as given by

$$f(x) = \sin(2\pi x/L),$$

then the *second harmonic* is given by $\sin(4\pi x/L)$, the *third harmonic* is given by $\sin(6\pi x/L)$, and so on. We obtain a large class of interesting functions when harmonics are summed. Here, we shall examine curves that are mostly sums of a few sine and cosine functions. (Chapter 13 will discuss sums of a large number of terms – the Fourier series.) In the introduction to this section, we showed that the cosine function is a special case of the general form $\sin(ax + b)$. Therefore, we can use only sine terms in the sums with a suitable phase b (or alternatively, use only cosine terms in the sums with a suitable phase).

First consider the sum of a fundamental sine wave with its second harmonic:

$$f(x) = \sin(2\pi x) + c\sin(4\pi x + b)$$

The graphs of this function for $c = 1$ and $b = n\pi/4$ ($n = 0, 1, 2, \ldots, 7$) are shown in Figure 5.2.9 and 5.2.10. Only those functions with $b = 0$ and $b = \pi$ have symmetry, and both are odd functions. There is further symmetry among these functions and can be expressed as

$$\sin(2\pi x) + \sin(4\pi x + n\pi/4) = -\sin(-2\pi x) - \sin(-4\pi x - n\pi/4)$$

Figures 5.2.11 and 5.2.12 show similar graphs for the cosine plus its second harmonic. There are just two of the functions that are symmetric ($b = 0$ and π), and their symmetry is even. Further symmetry exists among these functions and can be expressed as

$$\cos(2\pi x) + \cos(4\pi x + n\pi/4) = \cos(-2\pi x) + \cos(-4\pi x - n\pi/4)$$

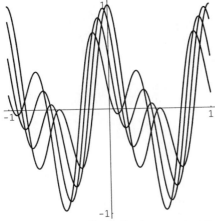

FIGURE 5.2.11.   Graph of cosine wave plus second harmonic ($b = 0$, $\pi/4$, $\pi/2$, $3\pi/4$).

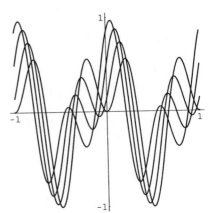

FIGURE 5.2.12.   Graph of cosine wave plus second harmonic ($b = \pi$, $5\pi/4$, $3\pi/2$, $7\pi/4$).

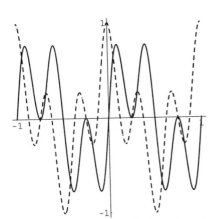

FIGURE 5.2.13.   Graphs of sine and cosine waves plus third harmonic ($b = 0$).

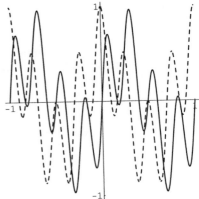

FIGURE 5.2.14.   Graphs of sine and cosine waves plus fourth harmonic ($b = 0$).

If we wish to consider higher harmonics, let us write the functions

$$f(x) = \sin(2\pi x) + c\sin(n\,2\pi x + b)$$
$$g(x) = \cos(2\pi x) + c\cos(n\,2\pi x + b)$$

where $n = 2, 3, \ldots$ and b is the arbitrary phase. Figures 5.2.13 and 5.2.14 show these functions for the values of $n = 3$ and $n = 4$ when $c = 1$ and $b = 0$.

In Chapter 13, we will explore in more detail the sums of harmonics and show that arbitrarily complex curves can be expressed as the sum of many harmonics. This will be done in the context of the Fourier series representation.

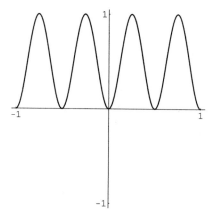

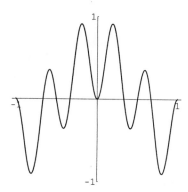

FIGURE 5.2.15. Graph of $\sin(2\pi x)\sin(2\pi x)$.   **FIGURE 5.2.16.** Graph of $\sin(2\pi x)\sin(3\pi x)$.

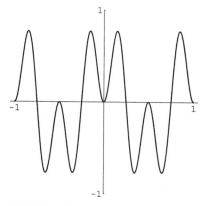

FIGURE 5.2.17. Graph of $\sin(2\pi x)\sin(4\pi x)$.

5.2.3. Products of Harmonics

A product of harmonics is expressed as one of the following:

1) $y = \sin(2\pi ax)\sin(2\pi bx)$
2) $y = \sin(2\pi ax)\cos(2\pi bx)$
3) $y = \cos(2\pi ax)\cos(2\pi bx)$
4) $y = \cos(2\pi ax)\sin(2\pi bx)$

The graphs of (1) and (3) are identical except for a phase shift of $\pi/2$; this is also the case for (2) and (4).

The graphs of (1) for $a = 1$ and $b = 1$, 3/2, 2 are shown in Figures 5.2.15 to 5.2.17, respectively. The curves increase in complexity as b increases, but there are predictable attributes of the curves which relate to values of b. These attributes, presuming the form of Case (1) above, are summarized as

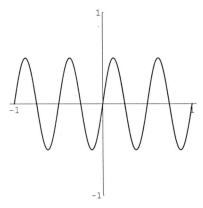

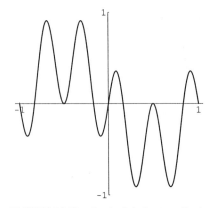

FIGURE 5.2.18. Graph of $\sin(2\pi x)\cos(2\pi x)$.   **FIGURE 5.2.19.** Graph of $\sin(2\pi x)\cos(3\pi x)$.

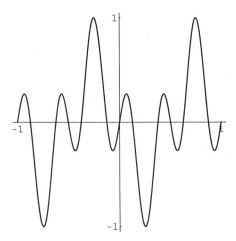

FIGURE 5.2.20. Graph of $\sin(2\pi x)\cos(4\pi x)$.

b	Fundamental period L	Number of peaks per L
Half integer (n/m)	2	$n + m$
Even integer n	1	$n + 1$
Odd integer n	1/2	$(n + 1)/2$

All of the curves of Case (1) have even symmetry due to the fact that they are the product of two functions with odd symmetry.

Now consider the product of harmonics given by Case (2) above. The graphs of (2) for $a = 1$ and $b = 1, 3/2, 2$ are shown in Figures 5.2.18 to 5.2.20, respectively. The relations between certain properties of the sin $*$ cos curves and the value of b are the same as those listed above for the sin $*$ sin functions. However, the sin $*$ cos curves all show odd symmetry due to the fact that they are the product of an even and an odd function.

5.3. CYCLOIDS

5.3.1. Parametric Representation

Thus far in this chapter, we have studied curves that are single-valued on x. In order to construct more complicated and interesting curves, we need a means of making multivalued curves on x. The parametric form of a function is such a means. Recall that a parametric curve on the plane can be expressed with

$$x = f(t)$$
$$y = g(t)$$

When $f(t)$ has a periodic component in it, then x may, but not necessarily, take on the same value more than once as t varies. When $g(t)$ is periodic at the same time, the effect is to make curves which often cross over themselves. We will see many of these in this section.

5.3.2. Cycloids

Cycloids are a family of curves that have a simple physical basis which involves a rolling wheel. Consider the wheel of radius a supported by the x axis with its center directly over the origin $(0, 0)$. Let an arm of length b extend downward from the center of the wheel and be affixed to the wheel. Now let the wheel roll along the x axis; the curve traced by the end of the arm is called a *cycloid*. Depending on the ratio of b/a, the cycloid can be classified as

1. *ordinary cycloid* for $b = a$
2. *prolate cycloid* for $b > a$
3. *curtate cycloid* for $b < a$

The parametric equations for the cycloid as described above are

$$x = at - b\sin(t)$$
$$y = a - b\cos(t)$$

It is sometimes convenient to express y as simply $b\cos(t)$. This only causes a downward translation of the curve and a reversal in sign but does not affect its shape. The fundamental period of the cycloid, as written here, is 2π.

Figures 5.3.1 to 5.3.3 show the ordinary cycloid, a prolate cycloid, and a curtate cycloid, respectively, for $a = 1$. Note that using $b > a$ (prolate) will cause loops to appear in the curve and using $b < a$ (curtate) makes them disappear. The case $a = b$ is the borderline case and results in cusps that have a discontinuity in derivative of the curve.

5.3.3. Variants of Cycloids

We can make variants of the cycloid by letting

$$x = at - b\sin(t)$$
$$y = d\cos(t)$$

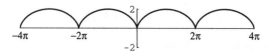

FIGURE 5.3.1. Ordinary cycloid ($b/a = 1$).

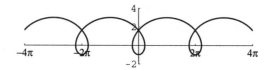

FIGURE 5.3.2. Prolate cycloid ($b/a = 2$).

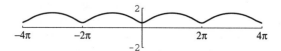

FIGURE 5.3.3. Curtate cycloid ($b/a = 1/2$).

FIGURE 5.3.4. A variation of the cycloid ($d/b = 2$).

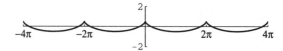

FIGURE 5.3.5. A variation of the cycloid ($d/b = 1/2$).

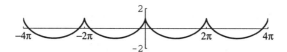

FIGURE 5.3.6. A variation of the cycloid ($d/b = 1$).

where d differs from b. The effect is as though the length of the arm varies during the cycle in an elliptical manner. Using $a = b$, Figures 5.3.4 to 5.3.6 show cases where $d = 2b$, $d = b/2$, and $d = b$ (ordinary cycloid for comparison), respectively.

Another, more interesting, variant of the cycloid can be made by letting

$$x = at - b\sin(t)$$
$$y = b\cos(mt/n)$$

where m and n are integers. This has the effect of making the periodicity in y shorter ($m > n$) or longer ($m < n$) than that of the periodic term in x. The fundamental period of this variant of the cycloid is equal to the longer of the two periods: 2π or $(n/m)2\pi$. We can generate a large range of interesting curves from this variant form. The two

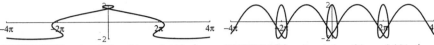

FIGURE 5.3.7. A variation of the cycloid ($m/n =$ 1/3 in argument for cosine).

FIGURE 5.3.8. A variation of the cycloid ($m/n =$ 2 in argument for cosine).

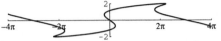

FIGURE 5.3.9. A variation of the cycloid ($m/n =$ 1/3 in argument for sine).

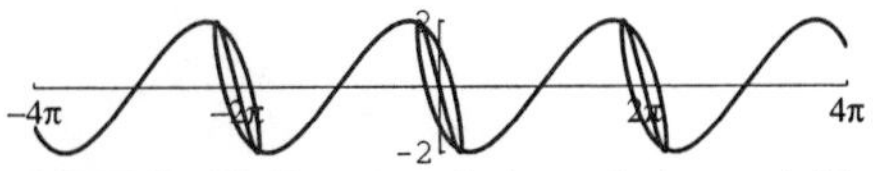

FIGURE 5.3.10. A variation of the cycloid ($m/n = 2$ in argument for sine).

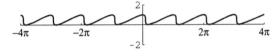

FIGURE 5.3.11. A variation of the cycloid (power = 2).

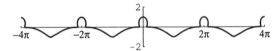

FIGURE 5.3.12. A variation of the cycloid (power = 3).

FIGURE 5.3.13. A variation of the cycloid (power = 4).

examples in Figures 5.3.7 and 5.3.8 are variations on the prolate cycloid using $a = 1$ and $b = 2$, respectively.

Another variant substitutes sin for cos in the y expression, thus

$$x = at - b\sin(t)$$
$$y = b\sin(mt/n)$$

When this substitution is made, the plots in Figures 5.3.7 and 5.3.8 become those shown in Figures 5.3.9 and 5.3.10, respectively. Note the symmetry of the former two and the antisymmetry of the latter two.

A final variant of the cycloid involves using an arbitrary power on the trigonometric functions thus

$$x = at - b\sin^n(t)$$
$$y = b\cos^n(t)$$

Examples of this variant using $n = 2, 3, 4$ and $a = b = 1$ are shown in Figures 5.3.11 to 5.3.13, respectively. Note that even powers of n will result in positive values only for y and that only odd powers of n produce a graph symmetric about the y axis.

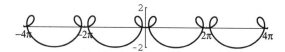

FIGURE 5.4.1. Compound cycloid ($a = b = c(+-) = 1, d = 2$).

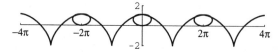

FIGURE 5.4.2. Compound cycloid ($a = b = c(--) = 1, d = 2$).

FIGURE 5.4.3. Compound cycloid ($a = b = c(++) = 1, d = 2$).

FIGURE 5.4.4. Compound cycloid ($a = b = c(-+) = 1, d = 2$).

5.4. COMPLEX PARAMETRIC CURVES ON A LINE

5.4.1. Compound Cycloid

Recalling the cycloid above, let us suppose that another wheel of radius c is attached (at its center) to the end of the arm of length b. Let this wheel maintain an independent rotation rate that is some multiple or submultiple of the rate of the main wheel. We shall call such an arrangement a *compound cycloid*. Let us examine the trace of a point on the perimeter of the second wheel ; its functional expression is

$$x = at - b\sin(t) \pm c\sin(dt)$$
$$y = b\cos(t) \pm c\cos(dt)$$

The plus or minus sign in the expressions allow for those cases where the point on the second wheel may start on the top or the bottom and where this wheel itself may rotate clockwise or counterclockwise. Four distinct curves are thus possible. The additional terms for x and y lead to intricate curves having patterns of multiple loops.

Figures 5.4.1 to 5.4.4 show four examples, all using $a = b = c = 1$ and $d = 2$, with the four permutations of the plus and minus signs on the constant c in the above equations. The signs in the annotation following "c" indicate the signs used for the x and y components, respectively. Physically, the second wheel overlays the first but has a rotation rate twice as fast as that of the first.

Figure 5.4.5 shows a realization with $a = b = c = 1$ and $d = 3$ with $+c$ in both the x expression and the y expression. Note that the complexity of the curve here has increased compared with the curves for $d = 2$ above. Increasing d further will

FIGURE 5.4.5. Compound cycloid ($a = b = c(++) = 1, d = 3$).

FIGURE 5.4.6. Compound cycloid ($a = b = c(++) = 1, d = 5/2$).

FIGURE 5.4.7. Generalized compound cycloid example.

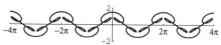

FIGURE 5.4.8. Generalized compound cycloid example.

FIGURE 5.4.9. Generalized compound cycloid example.

FIGURE 5.4.10. Generalized compound cycloid example.

steadily increase the complexity of the curve because the second wheel is making multiple revolutions per single revolution of the first wheel.

The period of the compound cycloids thus far is 2π. If d is not an integer but a fraction m/n, the period of the curve is altered to $2\pi n$. Figure 5.4.6 shows the compound cycloid of Figure 5.4.5, but with d changed to 5/2.

5.4.2. Generalized Compound Cycloid

The compound cycloid is a special case of the more general form

$$x = at + b_1 \sin(kt) + c_1 \sin(lt)$$
$$y = b_2 \cos(mt) + c_2 \cos(nt)$$

where the coefficients b_1, c_1, b_2, and c_2 are arbitrary and the factors k, l, m, and n are also arbitrary. We can't explore all the possible variations when using the general form, but we will look at some particular realizations. Consider, for instance, the following:

$$x = at + b \sin(kt)$$
$$y = c \cos(mt) + d \cos(nt)$$

This curve is symmetric with respect to the y axis and has a repeat period equal to the longer of $2\pi/k$, $2\pi/m$, or $2\pi/n$. For the curves of this section, we will take m to be equal or less than the smaller of k or n. Figure 5.4.7 shows the curve for these particular values: $a = 1$, $b = 1$, $c = 1$, $d = 1/2$, $k = 5$, $m = 1$, and $n = 4$. This example does not have similar forms above and below the x axis. We can make it have similar forms by making the constant n an odd integer, say 5, and the constant k an even integer, say 4. The result is shown in Figure 5.4.8.

If k and n are both odd or both even, we again obtain a curve that has dissimilar forms above and below the x axis (see the two examples of Figures 5.4.9 and 5.4.10). The first uses $k = n = 5$ and the second uses $k = n = 4$.

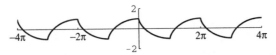

FIGURE 5.4.11.　Cycloid with $|\sin(t)|$.

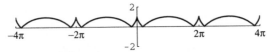

FIGURE 5.4.12.　Cycloid with $|\cos(t)|$.

FIGURE 5.4.13.　Cycloid with both $|\sin(t)|$ and $|\cos(t)|$.

The complexity of these curves is governed by the ratio of k to m and n to m. Larger values of m or n produce more vertical oscillation while larger values of k produce more horizontal oscillation.

5.4.3.　Forms Using Absolute Value of Trigonometric Functions

Ignoring the constant term in the y expression, the parametric form for the cycloid is

$$x = at - b\sin(t)$$
$$y = b\cos(t)$$

We can see some interesting effects when the absolute value of the periodic term for either x or y is used. This is expressed as

$$x = at - b|\sin(t)|$$
$$y = b|\cos(t)|$$

Figure 5.4.11 uses the absolute value of $\cos(t)$, but not of $\cos(t)$, with $a = b = 1$. Compare this with the ordinary cycloid in Figure 5.3.6. There is no longer any symmetry about the y axis, and the current graph is an example of an odd-harmonic function which has $f(x + L/2) = -f(x)$ when L is the fundamental period. Figure 5.4.12 uses the absolute value of $\cos(t)$, but not of $\sin(t)$, again with $a = b = 1$. Symmetry about the y axis is retained. Lastly, Figure 5.4.13 uses the absolute value of both $\sin(t)$ and $\cos(t)$, again with $a = b = 1$. All values are positive, the result of flipping the negative y parts of the first curve in this section up to positive y.

We can apply this effect to any of the variants of the cycloid or other parametric forms given thus far. The effect is somewhat predictable—after applying the absolute–value operation, the periodic term, in either x or y, will have only positive values. This operation produces curves with discontinuities in derivatives at points where it was previously smooth. Another operation that will produce effects similar to that of taking the absolute value is to take the square of either of the trigonometric terms in the expressions above.

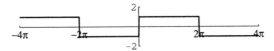

FIGURE 5.4.14. Square sine wave with period 4π.

FIGURE 5.4.15. Square cosine wave with period 4π.

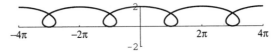

FIGURE 5.4.16. Cycloid variant before square-wave multiplication.

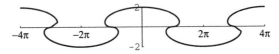

FIGURE 5.4.17. Cycloid variant after square-wave multiplication.

5.4.4. Transformations Using a Square Wave

Another interesting operation is to apply the square wave to a cycloid or one of its variants. A square sine wave or cosine wave of period L is defined as

$$(-1)^{\text{INT}(2x/L)} \text{ (sine)}$$
$$(-1)^{\text{INT}[2(x+L/4)/L]} \text{ (cosine)}$$

where "INT" means to take the integer part of the argument in brackets and discard any fractional part. Figures 5.4.14 and 5.4.15 show a square sine wave and a square cosine wave, respectively, each with a period of 4π.

As an example of the effect of applying a square wave as a factor, we consider this variant of a cycloid:

$$x = t + 2\sin(t)$$
$$y = 1 + \cos(t)$$

which is plotted in Figure 5.4.16. The curve was purposely meant to be tangent to the x axis; this is required to produce the desired effect. The fundamental period for this curve is 2π. In order to reverse the sign of y every other period, the y function is multiplied by a square cosine wave of period 4π. The result is shown in Figure 5.4.17. The curve remains continuous (and also the first derivative) after this transformation because the change in sign comes exactly at points of zero derivative (tangency to the x axis).

A second example takes the curve given by

$$x = t + \sin(4t)$$
$$y = 1 - \cos(4t)$$

FIGURE 5.4.18. Cycloid variant before square-wave multiplication.

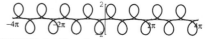

FIGURE 5.4.19. Cycloid variant after square-wave multiplication.

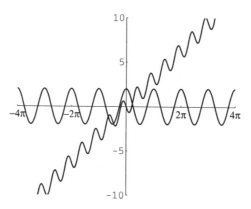

FIGURE 5.5.1. Graphs of $x = t - \sin(4t)$ and $y = 2\cos(2t)$.

This is plotted in Figure 5.4.18. The repeat period for this curve is $\pi/2$. To reverse the sign of y every other period, the y function is multiplied by a square sine wave of period π. The result is shown in Figure 5.4.19.

This sign-change operation can produce interesting and useful curves which have the same form above and below the x axis. Note that this is not the same as symmetry about the x axis though.

5.5. ANALYSIS OF THE LOOPS OF PERIODIC FUNCTIONS ON A LINE

We have seen a wide variety of curves generated by the cyloid and its variants. Except for the simple cycloid, it is difficult to predict how these complex parametric functions will appear when plotted. It is often helpful to consider the two parts, x and y, before plotting the final function. We can change parameters until the x or y behavior appears as desired, then we can plot the parametric curve. Consider the following cycloid variant:

$$x = t - \sin(4t)$$
$$y = 2\cos(2t)$$

We may want to know whether the loops overlap in x. Let us first plot x versus t and y versus t as in Figure 5.5.1. The graph of x versus t shows that for certain ranges of x three values of t will give an x in that range. This is indicative of a loop, but do they overlap when x and y are plotted as a parametric function? The answer lies in looking at the minima and maxima of x versus t in conjunction with the period of y. We can see from the graphs of Figure 5.5.1 that the basic repeat period of the parametric plot

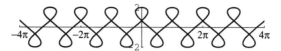

FIGURE 5.5.2. Cycloid variant with non-overlapping loops.

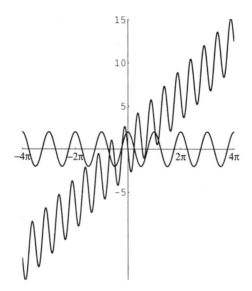

FIGURE 5.5.3. Graphs of $x = t - 3\sin(4t)$ and $y = 2\cos(2t)$.

will be that of y versus t, which is 2π. Successive loops of the parametric curve are on the positive and negative sides of the x axis. In the present case, the maximum associated with one loop is not greater than the minimum associated with the next loop at the period of 2π (two loops away); therefore, the loops will not overlap. By plotting the above parametric equations as one graph in Figure 5.5.2, we can see that these inferences are correct.

If we want the loops to overlap, we should increase the coefficient on the sine term in the x component thus:

$$x = t - 3\sin(4t)$$
$$y = 2\cos(2t)$$

Plotting these functions separately versus t gives the graphs shown in Figure 5.5.3. The x versus t curve has the desired overlap between the maximum and minimum of loops on the same side of the x axis. The resulting graph in Figure 5.5.4 shows that the loops do indeed overlap. If we halve the period of the y oscillation by using $y = 2\cos(4t)$, the result can change greatly (see Figure 5.5.5).

These few examples illustrate that variants of the cycloid form can be analyzed and controlled such that a craftful approach to using them can produce a desired curve shape.

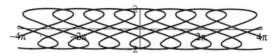

FIGURE 5.5.4. Cycloid variant with overlapping loops.

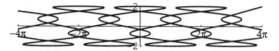

FIGURE 5.5.5. Cycloid variant with overlapping loops.

5.6. SUMMARY

Periodic functions on a line have $f(x + L) = f(x)$ where L is the fundamental period. The fundamental functions of this kind are the sine and cosine function. Interesting variants can be generated by taking sums or products of harmonics; the results also being harmonic. When parametric forms are used, the complexity of periodic functions on the line can be increased greatly. Cycloids are the prototypical examples of the parametric form. Variations on cycloids can be made with additional harmonic terms. Such variations can usually be done in a predictable manner. Although these curves were all generated on a line ($x \propto t$), the parametric form can be easily modified to produce curves on a parabola ($x \propto t^{1/2}$), on a cubic ($x \propto t^{1/3}$), and so forth.

Chapter 6

PERIODIC FUNCTIONS ON A CIRCLE

6.1.   PERIODICITY AND SYMMETRY
FOR CIRCULAR FUNCTIONS

Periodic functions on a circle are very similar to the functions of the preceding chapter, except that the baseline of the function is along the circumference of a circle rather than along the x axis. Thus, functions that are periodic on the x axis can be "wrapped" around a circle to form periodic functions on a circle. Therefore, instead of

$$y = f(x)$$

or, for parametric functions,

$$x = f(t)$$
$$y = g(t)$$

we have

$$r = f(\theta)$$

or, for parametric functions,

$$\theta = f(t)$$
$$r = g(t)$$

where r is the distance from the origin and θ is the angle counterclockwise from the positive x axis. We use the polar coordinates of Chapter 1 rather than the Cartesian coordinates. Formerly, the limits of x were infinite; however, the limits of θ will be from zero to some multiple (or possibly submultiple) of 2π. If, as is the case for all functions to be considered here, we require that the endpoints of the function meet, then certain constraints are placed on the coefficients of the function. (No such constraints were placed on periodic functions on a line.) The periodicity of a function on the circle can be stated in terms of the angle θ:

$$r(\theta + 2\pi/n) = r(\theta)$$

where n is some integer number equal to two or greater. For the limiting case of $n = 2$, the curve will have two *lobes* that are identical forms rotated along the circumference by an angle of π. Indeed, for any value of n, the number of lobes is equal to n, each shifted by an angle equal to $2\pi/n$. We have excluded curves for which $n = 1$. These curves show only one complete cycle for θ in the range 0 to 2π and will be covered in a later chapter.

In regard to symmetry for functions on a circle, we are first interested in the *point symmetry* of the curve. The "point" is the origin $(0, 0)$. A curve is point symmetric

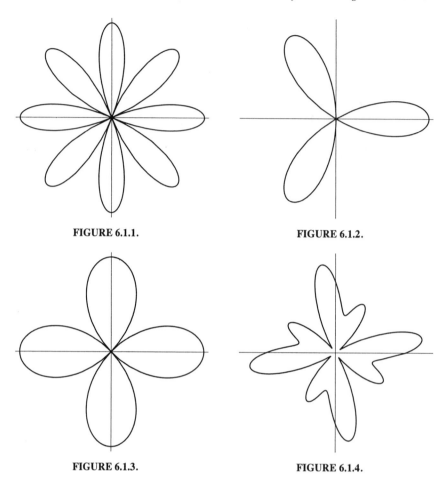

FIGURE 6.1.1.

FIGURE 6.1.2.

FIGURE 6.1.3.

FIGURE 6.1.4.

if $r(\theta + \pi) = r(\theta)$ for all θ. The periodic curve with $n \geq 2$ will show such symmetry whenever $n = 2, 4, 6,...$ (an even number of lobes). If $n = 3, 5, 7, \ldots$ (an odd number of lobes), then the function will not show point symmetry. Figures 6.1.1 and 6.1.2 illustrate the presence or absence of point symmetry, respectively.

In the case of curves in Cartesian coordinates, we considered symmetry about the x and y axes. But for curves periodic in θ, we consider symmetry about any arbitrary line through $(0, 0)$. Thus, lines of symmetry pass through the origin and divide the curve into mirror images. We first look at the curves with an even number of lobes. They fall into one of two classes: 1) those with n lines of symmetry and 2) those with no lines of symmetry. If there are any lines of symmetry, their number will exactly equal the number of lobes and will occur at angles π/n apart. In addition, such curves will have an infinite number of lines of antisymmetry. Such lines pass through the origin and divide the curve into two parts: one of which must be rotated by π to overlap the other. The curve in Figure 6.1.3 has point symmetry. This curve also has four lines of symmetry: at $\theta = 0, \pi/4, \pi/2$, and $3\pi/4$. However, Figure 6.1.4

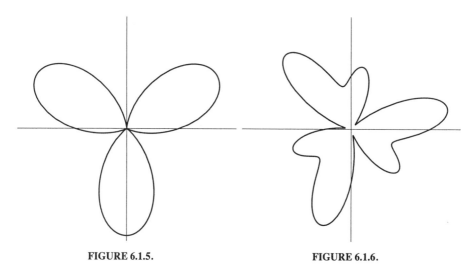

FIGURE 6.1.5.   **FIGURE 6.1.6.**

again shows a curve with four lobes but no lines of symmetry even though it is point symmetric.

Let us now consider the curves with an odd number of lobes. Recall that these curves have no point symmetry. They also fall into two classes: 1) those having a number of lines of symmetry equal to the number of lobes and 2) those with no lines of symmetry. In the former class, we placed these lines of symmetry $2\pi/n$ apart in θ where n is the number of lobes. Figures 6.1.5 and 6.1.6 illustrate the two classes.

6.2. SIMPLE PERIODIC FUNCTIONS

6.2.1. Basic Harmonics

The first class of periodic curves on the circle to be considered are those for which

$$r = a + b\cos(m\theta + \phi)$$

where m is an integer equal to 2 or greater. Note that ϕ is an arbitrary phase angle that has the effect of merely rotating the curve; therefore, because it has no real effect on the shape of the curve, it will be dropped in all of the examples. By letting $\phi = \pi/2$, for instance, the "cos" function would be changed to "sin" and the curve would be rotated by $\pi/2$. Figure 6.2.1 shows the function given above for $m = 2, 3, 4$ and $a = 3/4, b = 1/4$. Note that there are m lobes when m is either an even or odd integer. Figure 6.2.2 shows the same values of m as Figure 6.2.1 but with $a = 1/2, b = 1/2$. Note that, when $a = b$, the lobes all approach zero width and meet at the origin. Figure 6.2.3 uses the same values of m again but with $a = 1/4, b = 3/4$. Note that the number of lobes is doubled. This lobe doubling occurs whenever $b > a$.

The properties of the function $r = a + b\cos(m\theta)$ can be summarized by the following

$a > b \rightarrow$ lobes are identical; no point of curve passes through origin; m lobes for even m and also m lobes for odd m

$a < b \rightarrow$ two sets of m identical lobes each; curve passes through origin $2m$

FIGURE 6.2.1. Graphs of $3/4 + \cos(m\theta)/4$ for $m = 2, 3, 4$.

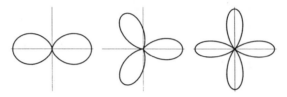

FIGURE 6.2.2. Graphs of $1/2 + \cos(m\theta)/2$ for $m = 2, 34$.

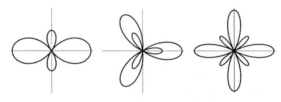

FIGURE 6.2.3. Graphs of $1/4 + 3\cos(m\theta)/4$ for $m = 2, 3, 4$.

FIGURE 6.2.4. Graph of $\cos(m\theta)$ for $m = 2, 3, 4$.

times for even m and also $2m$ times for odd m; $2m$ lobes for even m and also $2m$ lobes for odd m

$a = b \rightarrow$ lobes are identical; curve passes through origin m times for even or odd m; m lobes for even or odd m

An interesting group of the above curves, $a + b\cos(m\theta)$, is formed when $a = 0$. They are called "rose" curves or "rhodonea." Note that r becomes negative for certain ranges of θ; this is handled by a reflection of the point at $(|r|, \theta)$ through the origin to $(|r|, \theta + \pi)$. Replotting the above set of functions with $a = 0$ and $b = 1$ provides some examples of the rhodonea shown in Figure 6.2.4. Note that for odd m the number of lobes is halved. This is due to the coincidence of the graph for $0 < \theta < \pi$ and $\pi < \theta < 2\pi$. Therefore, it is only necessary to use half of the range of θ in the case

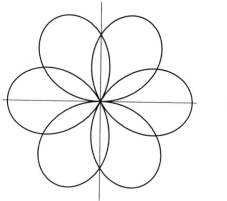

FIGURE 6.2.5. Graph of cos(3θ/2).

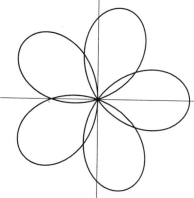

FIGURE 6.2.6. Graph of cos(5θ/3).

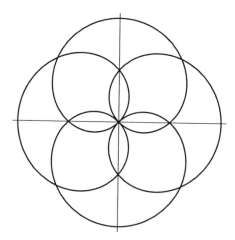

FIGURE 6.2.7. Graph of cos(2θ/3).

of odd *m* when *a* = 0. A family of curves which generalizes the rhodonea is given by

$$r = \cos(m\theta/n)$$

where *m/n* is rational. A set of three examples is shown in Figures 6.2.5 to 6.2.7. These curves may seem somewhat unpredictable, but the expected results can be exactly predicted according to the integers *m* and *n* as tabulated here

$$m \text{ odd }, n \text{ even:} \quad 2m \text{ lobes }, 0 < \theta < 2n\pi$$
$$m \text{ odd }, n \text{ odd:} \quad m \text{ lobes }, 0 < \theta < n\pi$$
$$m \text{ even }, n \text{ odd:} \quad 2m \text{ lobes }, 0 < \theta < 2n\pi$$

We can see that the examples plotted above follow this scheme. We can add a constant to these plots by using

$$r = a + b\cos(m\theta/n)$$

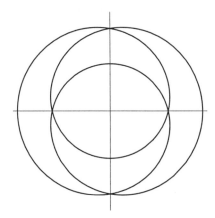

FIGURE 6.2.8. Graph of $3/4 + \cos(2\theta/3)/4$.

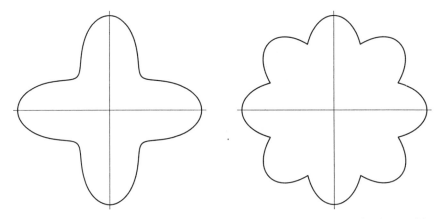

FIGURE 6.2.9. Graph of $3/4 + \cos(4\theta)/4$.   **FIGURE 6.2.10.** Graph of $3/4 + |\cos(4\theta)|/4$.

One example of using this form is plotted in Figure 6.2.8 for $a = 3/4, b = 1/4, m = 2$, and $n = 3$. Interesting modifications to $r = a + b\cos(m\theta)$ are

$$r = a + b|\cos(m\theta)|$$

and

$$r = a + b\cos^2(m\theta).$$

Consider the function $r = 3/4 + \cos(4\theta)/4$ plotted in Figure 6.2.9. Now we plot the same equation but with the absolute-value operation performed on the cosine function. The result is shown in Figure 6.2.10. The absolute-value operation causes the cusps to appear where the first derivative of r with respect to θ is discontinuous. Note that the number of lobes has been doubled too because the absolute-value operation makes twice as many peaks in the cosine function.

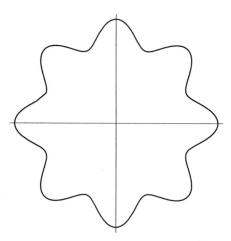

FIGURE 6.2.11. Graph of $3/4 + [\cos(4\theta)]^2/4$.

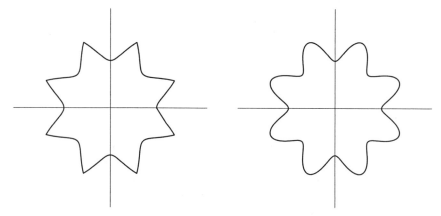

FIGURE 6.2.12. Graph of $3/4 - |\cos(4\theta)|/4$.   **FIGURE 6.2.13.** Graph of $3/4 - [\cos(4\theta)]^2/4$.

Figure 6.2.11 shows the same equation but with a squaring operation rather than an absolute-value operation. The squaring operation, like the absolute-value operation, doubles the number of lobes; but it does not produce discontinuities in the first derivative of r. The reason it does not, while the absolute-value operation does, relates to the behavior of the cosine function as it approaches zero. In this vicinity, $\cos(\theta) \sim \theta$; therefore, $\cos^2(\theta) \sim \theta^2$, the derivative of which is continuous. In the case of the absolute-value operation, the derivative of $|\cos(\theta)|$ as it approaches zero is a constant which abruptly changes sign when $\cos(\theta) = 0$, thus producing the cusps.

The shapes of these curves can be inverted by simply changing the sign of the coefficient b. We do this first for the absolute-value operation in Figure 6.2.12 and then for the squaring operation in Figure 6.2.13.

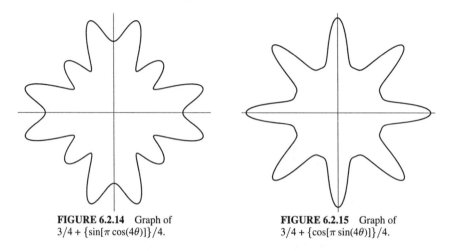

FIGURE 6.2.14   Graph of
$3/4 + \{\sin[\pi \cos(4\theta)]\}/4$.

FIGURE 6.2.15   Graph of
$3/4 + \{\cos[\pi \sin(4\theta)]\}/4$.

6.2.2.   Compound Harmonics

We express a periodic function on the circle using a compound harmonic as one of the following

$$1) \quad r = a + b\sin[m\pi\sin(2\pi\theta)]$$
$$2) \quad r = a + b\sin[m\pi\cos(2\pi\theta)]$$
$$3) \quad r = a + b\cos[m\pi\cos(2\pi\theta)]$$
$$4) \quad r = a + b\cos[m\pi\sin(2\pi\theta)]$$

Similar functions were introduced in Section 5.2.1 as periodic functions on a line. There, we pointed out that (1) and (2) are identical except for a phase shift; this was true for (3) and (4) also. Therefore, we will consider only (2) and (4) here. Their behavior is illustrated with the examples of Figures 6.2.14 and 6.2.15.

The variety of shapes possible with the compound harmonics is too large to provide a full suite of examples here. As for the simple harmonic, a value of $b > a$ produces a new and more complex shape. Figure 6.2.16 shows one example using $a = 1/4$ and $b = 3/4$ for the compound harmonic.

6.2.3.   Sums of Harmonics

Sums of harmonics were introduced in Section 5.2.2 as periodic functions on the line. We will use these again, adapted to the circle. The periodic function on the circle using a fundamental harmonic and a single higher harmonic is

$$r = a + b\cos(m\theta) + c\cos(nm\theta + f)$$

where n and m are integers. If $\phi = 0$ and b and c are both positive, every other peak of the second harmonic is coincident with a peak of the fundamental harmonic and the maximum of the combined function has an amplitude of $a + b + c$ at these points. In this case, the range of the two harmonic terms is $[s, b+c]$ where s is to be determined by finding the minimum value by algebraic means. Then we can classify the curves

FIGURE 6.2.16. Graph of $1/4 + 3\{\cos[\pi \sin(2\theta)]\}/4$.

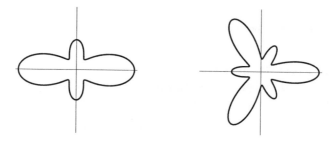

FIGURE 6.2.17. Graphs of $1/2 + [\cos(m\theta) + \cos(nm\theta)]/4$ for $n = 2, m = 2, 3$.

according to

$$a > -s$$
$$a = -s$$
$$a < -s$$

similar to the classification of the basic harmonics in 5.2.1. We start with two examples of the case where $a > -s$ as shown in Figure 6.2.17. For this function, $s \sim -0.28125$.

The next pair of examples, shown in Figure 6.2.18, are for the case where $a = -s$. The function touches the origin between lobes, as did the basic harmonic alone when a was set to the negative of the minimum of the harmonic function.

Figure 6.2.19 illustrates the case of a $< -s$ by simply letting $a = 0$. When the curves pass through the origin as in this last pair of the examples, the number of loops is not doubled as was the case for the simple harmonic in Figure 6.2.1. Here the ratio of loops is $3/2$, which is a consequence of the higher complexity of the function created from summing harmonics.

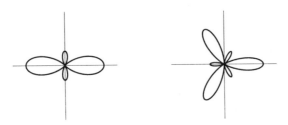

FIGURE 6.2.18.   Graphs of $0.28125 + [\cos(m\theta) + \cos(nm\theta)]/4$ for $n = 2, m = 2, 3$.

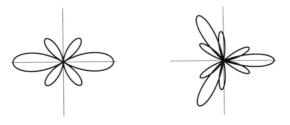

FIGURE 6.2.19.   Graphs of $[\cos(m\theta) + \cos(nm\theta)]/2$ for $n = 2, m = 2, 3$.

6.2.4.   Products of Harmonics

Section 5.2.3 introduced products of harmonics for periodic functions on a line. A similar suite of functions for periodic functions on a circle is

$$
\begin{aligned}
&1) \quad r = a + b\sin(m\theta)\sin(n\theta)\\
&2) \quad r = a + b\sin(m\theta)\cos(n\theta)\\
&3) \quad r = a + b\cos(m\theta)\cos(n\theta)\\
&4) \quad r = a + b\cos(m\theta)\sin(n\theta)
\end{aligned}
$$

We allow only ratios n/m which are rational. In Section 5.2.3, it was pointed out that (1) and (3) were identical to within a phase shift of $\pi/2$ and that (2) and (4) were also identical except for the same phase shift. This carries over when the functions are adapted to the circle; therefore, we will consider only the forms (3) and (4) here.

Let us look at some representative plots for Case (3) first: let m remain constant while varying n as illustrated in Figure 6.2.20. The use of Case (4) above with the same constants gives the curves shown in Figure 6.2.21. Note that the third function in Figure 6.2.21, involving $\cos(2\theta)\sin(6\theta)$ has no lines of symmetry although the function involving $\cos(2\theta)\cos(6\theta)$ does have them. In general, if the ratio n/m is an odd integer (except for 1) and the harmonic functions are not identical, then the resulting curve will show no planes of symmetry although it will have infinite planes of antisymmetry as discussed in Section 6.1.

As for the previous classes of periodic functions on the circle, the curves get more complex as they pass through the origin due to the constant a becoming small enough that negative r exists as θ varies. The exact value of a for which r becomes zero can be found by algebraic means, and then smaller values of a can be used to generate the more complex curves. An example is the function shown in Figure 6.2.22.

FIGURE 6.2.20. Graphs of $[1 + \cos(m\theta)\cos(nm\theta)]/2$ for $m = 2, n = 1, 2, 3$.

FIGURE 6.2.21. Graphs of $[1 + \cos(m\theta)\sin(nm\theta)]/2$ for $m = 2, n = 1, 2, 3$.

FIGURE 6.2.22. Graph of $[1 + 3\cos(\theta)\cos(2\theta)]/4$.

Many interesting and varied curves can be generated with the product of harmonics. Only a small sample of these has been shown here.

6.3. PERIODIC FUNCTIONS BASED ON POLYGONS

The *regular polygons* are defined as those polygons with equal angles at all vertices. This implies that all sides have equal length. In a sense, these are periodic functions on the circle because they have n repeated lobes for n sides. There are n lines of symmetry for an n-sided regular polygon when n is even; but, when n is odd, there are no lines of symmetry nor of antisymmetry. Although the most efficient way to plot polygons is to simply plot lines connecting the evenly spaced vertices around the center, it will be useful to have a functional relation that compactly describes a polygon as a periodic function on the circle. The analysis is not difficult for n general;

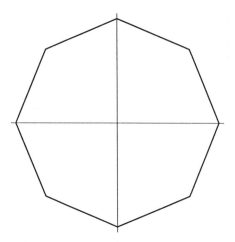

FIGURE 6.3.1.   Octagon plotted as a periodic function.

without proof, we give the desired function as

$$r = \sin\alpha/\sin(\pi - \alpha - \theta')$$

where

$$\alpha = (n - 2)\pi/(2n)$$
$$\theta' = \mathrm{mod}\,(\theta, 2\pi/n)$$

The "mod" operation produces the fractional remainder of dividing θ by the value $2\pi/n$. The value $2\pi/n$ should be recognized as the fundamental period of the function for there are n sides to the polygon as θ goes from 0 to 2π. Now that we have a functional relation $r = f(\theta)$ for polygons, we can make use of the same graphing method used for all the other periodic functions on the circle. The plot of the octagon (8 sides) is generated in this manner as shown in Figure 6.3.1.

With the above functional form for polygons, it becomes easy to modify them with functions having the same period and thereby produce new and interesting shapes. Let a modifying function be $g(\theta)$; then we will form the new function

$$r = a[\sin\alpha/\sin(\pi - \alpha - \theta')] + bg(\theta)$$

Let $g(\theta)$ be such that a harmonic of period $2\pi/n$ is added to the polygon's side; a form of $g(\theta)$ that will do this is

$$g(\theta) = \cos(n\theta),$$

Figure 6.3.2 shows the result when $a = 7/8$, $b = 1/8$, and $n = 4$; the square appears to be shrinking at the sides. By changing the sign of the constant b, we produce the expanding square seen in Figure 6.3.3.

If the modifying function is changed to

$$g(\theta) = \sin(n\theta),$$

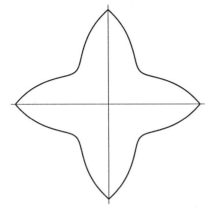

FIGURE 6.3.2. Square with $\cos(4\theta)$ added.　　**FIGURE 6.3.3.** Square with $-\cos(4\theta)$ added.

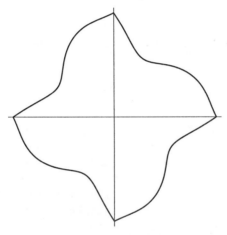

FIGURE 6.3.4. Square with $\sin(4\theta)$ added.

the phase of the modifying function is shifted by $\pi/(2n)$ for the polygon of n sides compared to the cosine function. The effect of using a sine function for $a = 1, b = 1/8$, and $n = 4$ is shown in Figure 6.3.4.

We can see that the sine weighting produces antisymmetry. If the sign of b is changed for the sine function, we obtain a graph that is the reflection of the above graph, either about the x axis or the y axis, as seen in Figure 6.3.5.

If the period of the weighting function is doubled, then the figure is compressed or stretched. Figure 6.3.6 starts with the square again but with the additive function equal to $\cos(2\theta)$ rather than $\cos(4\theta)$.

In summary, the cosine additive function will make the polygon appear to shrink or expand along its sides, depending on the sign of the cosine function; and the sine additive function will make the corners of the polygon appear to rotate either clockwise or counterclockwise, depending on the sign of the sine function. All the examples here, except Figure 6.3.6, have used one cycle of the cosine or sine function

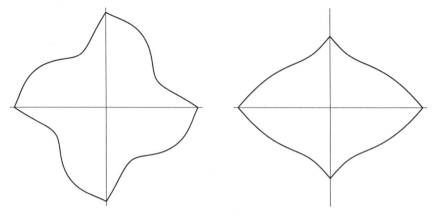

FIGURE 6.3.5. Square with $-\sin(4\theta)$ added.   **FIGURE 6.3.6.** Square with $\cos(2\theta)$ added.

per side of the polygon. More complex shapes are possible by using m cycles of these functions, where m is an integer, or by using compound harmonics, sums or harmonics, and products of harmonics.

6.4. TROCHOIDS

6.4.1. Hypotrochoids

The trochoids are the circular equivalent of the the cycloids studied in the previous chapter. Instead of the circle rolling on a line, it rolls around inside the circle (*hypotrochoids*) or outside the circle (*epitrochoids*) continuously keeping contact with the circle's circumference. We will consider the hypotrochoids first.

Figure 6.4.1 shows the geometry that gives rise to the hypotrochoid curve. Let the radius of the large circle be a, the radius of the small circle be b, and the length of the fixed arm extending from the center of the smaller circle be c. Let the position, as shown, be at $t = 0$. Then, as the smaller circle rolls around the circumference of the larger one, the trace of point P at the end of the arm will be given by

$$x = (a - b)\cos(t) + c\cos[(a - b)t/b]$$
$$y = (a - b)\sin(t) - c\sin[(a - b)t/b]$$

These equations are parametric in the variable t and can be computed as t varies from zero to some value to be determined (not necessarily 2π). If we choose to use polar plotting instead, we need r and θ in terms of t. The derivation from the above relations is straightforward and gives

$$r = \{(a - b)^2 + c^2 + 2c(a - b)\cos[at/b]\}^{1/2}$$
$$\theta = \tan^{-1}\{y/x\}$$

For computing the graph of the curve, the original x and y relations are clearly more efficient; and we will use them.

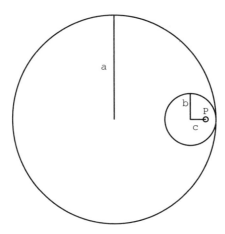

FIGURE 6.4.1. Hypotrochoid's geometry.

The constants a, b, and c are indicative of the form of the plotted curve. Because a/b is the ratio of the radii of the two circles, we require that it be rational in order for the curve to be able to connect with itself at the starting point. If a/b is a whole number, then this will occur as the smaller circle comes to its starting position after just one traverse of the circumference of the larger circle. When it is a whole number, the ratio a/b equals the number of lobes of the curve and the fundamental period of the curve equals 2π over the number of lobes, or $2\pi/(a/b)$.

If a/b is fractional, then we will need more traverses to make the curve connect. The exact number of traverses is given by n where n is the smallest integer multiplier that will make a/b into a whole number. For instance, if $a/b = 4/3$, then $n = 3$. Or, if $a/b = 10/8$, then $n = 4$. It follows that the range of t will be from 0 to $2\pi n$ in order to plot a complete hypotrochoid, that the number of lobes of the curve is given by na/b, and that the fundamental period of the curve is $2\pi/(na/b)$.

The quantity $a - b$ relative to c is also important in determining the shape of the curve. If a/b is a whole number, the overall shape will behave according to the following table

$a - b = c$ — curve passes through center of larger circle for each period
$a - b > c$ — curve does not go beyond the center for any period
$a - b < c$ — curve overshoots the center for each period

The ratio c/b controls the shape of the lobes. When $c > b$, the lobes show loops and increase in size as c increases. As c becomes very large compared to $a - b$, the curve approaches a circle of radius c. If $c = b$, the lobes collapse to cusps, and the curve is termed a *hypocycloid*. If $c < b$, the lobe effects become smaller as c decreases. The limit is for $c = 0$ where the plotted curve becomes a circle of radius $a - b$.

From this analysis of the hypotrochoid, it is evident that it can have a wide variety of forms within its definition. Some examples at this point will help to show the range of possibilities. The first suite of examples in Figure 6.4.2 shows three lobes and the

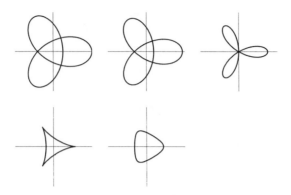

FIGURE 6.4.2. Hypotrochoids for $a = 9, b = 3, c = 10, 9, 6, 3, 1$.

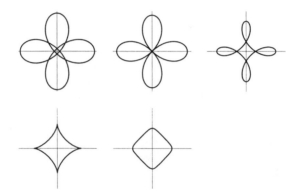

FIGURE 6.4.3. Hypotrochoids for $a = 12, b = 3, c = 10, 9, 6, 3, 1$.

second suite in Figure 6.4.3 shows four lobes with a/b equal to a whole number in both cases. The range of c values is such that all the possible forms are illustrated.

Figure 6.4.4 shows examples with $a/b = 9/6$; by the above analysis, this requires $n = 2$ so that the range of t is $[0, 4\pi]$ and the number of lobes is three. Figure 6.4.5 shows examples with $a/b = 5/3$; using the above analysis again, this requires $n = 3$ so that the range of t is $[0, 6\pi]$ and the number of lobes is five.

All of the above hypotrochoids had $a > b$. If $a = b$, then there can be no rolling motion of one circle relative to the other. But what about when $a < b$? This is also valid; for then the rolling circle can again move about the fixed circle, again requiring that the maximum t be $2\pi n$ in order to make a complete figure. All of the above discussion concerning the ratio a/b still applies when $a < b$. The number n is determined as before. For the quantity $a - b$, now substitute $b - a$ to determine the behavior with respect to c. Figure 6.4.6 provides examples with $a/b = 3/5$.

6.4.2. Epitrochoids
We now consider the epitrochoid. Figure 6.4.7 shows the geometry that gives rise to this curve. The definitions of a, b, and c match those of the hypotrochoid. Let the

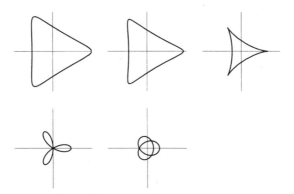

FIGURE 6.4.4. Hypotrochoids for $a = 9, b = 6, c = 10, 9, 6, 3, 1$.

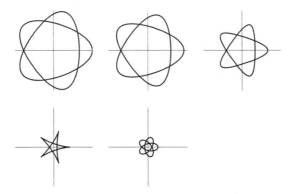

FIGURE 6.4.5. Hypotrochoids for $a = 5, b = 3, c = 10, 9, 6, 3, 1$.

FIGURE 6.4.6. Hypotrochoids for $a = 3, b = 5, c = 3, 2, 1$.

position shown be that at $t = 0$. Then, as the smaller circle rolls around the circumference of the larger one, the trace of point P at the end of the arm will be given by

$$x = (a + b) \cos(t) - c \cos[(a + b)t/b]$$
$$y = (a + b) \sin(t) - c \sin[(a + b)t/b]$$

The derivation of the polar plotting variables for the epitrochoid is again straightforward and gives

$$r = \{(a + b)^2 + c^2 - 2c(a + b) \cos[at/b]\}^{1/2}$$
$$\theta = \tan^{-1}\{y/x\}$$

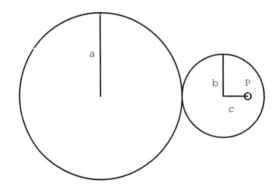

FIGURE 6.4.7. Epitrochoid's geometry.

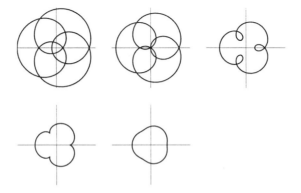

FIGURE 6.4.8. Epitrochoids for $a = 9, b = 3, c = 15, 12, 6, 3, 1$.

We will again use the more efficient (x, y) plotting variables though. The analysis of the epitrochoid leads to the same conclusions regarding the interplay of the variables a, b, and c as they did for the hypotrochoid. However, the quantity $a + b$, rather than $a - b$, must be used to determine the shape of the curve when a/b is a whole number according to the following table

$a + b = c$ — curve passes through center of larger circle for each period
$a + b > c$ — curve does not go beyond the center for any period
$a + b < c$ — curve overshoots the center for each period

Also note that when $b = c$ the form of the curve is then termed the *epicycloid*. We employ the same values of a, b, and c used for the hypotrochoid examples to illustrate the behavior of the epitrochoids. The two suites of curves in Figures 6.4.8 and 6.4.9 are for $a/b = 9/3$ and $a/b = 12/3$, respectively.

The case for a/b fractional is illustrated with $a/b = 9/6$ in Figure 6.4.10. Another case for a/b fractional is illustrated with $a/b = 5/3$ in Figure 6.4.11.

The above examples all had $a > b$. No real change is required for analyzing and predicting curve shapes when $a < b$. The ratio a/b still has the same implications,

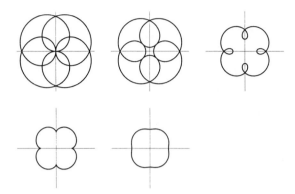

FIGURE 6.4.9. Epitrochoids for $a = 12, b = 3, c = 15, 12, 6, 3, 1$.

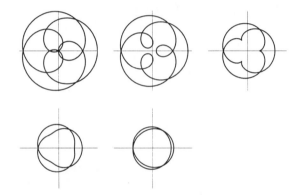

FIGURE 6.4.10. Epitrochoids for $a = 9, b = 6, c = 15, 12, 6, 3, 1$.

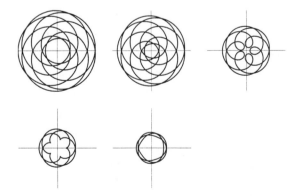

FIGURE 6.4.11. Epitrochoids for $a = 5, b = 3, c = 15, 12, 6, 3, 1$.

FIGURE 6.4.12. Epitrochoids for $a = 3, b = 5, c = 12, 8, 4$.

and the sum $a + b$ with respect to c still has the same effects. Figure 6.4.12 uses $a = 3$ and $b = 5$ to illustrate the case where $a < b$.

6.5. PARAMETRIC CURVES ON A CIRCLE

6.5.1. Parametric Representation in r and θ
This section will examine curves given by

$$r = a + b\cos(mt + \phi)$$
$$\theta = t + c\sin(nt)$$

Recall that the definition of a cycloid in the previous chapter was

$$y = b\cos(t)$$
$$x = at - b\sin(t)$$

The similarity between the present form for the circle and the cycloid leads us to term the curves "cycloid-like." The allowance for $m \neq n$ in the present case will, however, generate a wide variety of curves not possible with a strict carry-over of the cycloid to the circle. We will consider the case $m = n$ first though and then look at $m < n$ and $m > n$.

6.5.2. Curves with Period of r Equal to Period of θ
Let $m = n = 4$ in the above definitions for r and θ, and let $\phi = 0$ to produce the curve in Figure 6.5.1. We can see that the number of lobes is equal to the value of $m = n$. Figure 6.5.2 shows how the loops are turned outside by letting $\phi = \pi$ (equivalently changing the sign of the cosine term in the definition of r). If the phase ϕ is set at $\pi/2$, the curve appears to have a pinwheel shape as shown in Figure 6.5.3. A phase of $-\pi/2$ would simply reverse the direction of the arms.

In the above, the shape of the lobes can be controlled with the coefficients on the sine and cosine terms. Specifically, the angular range of the loops or arms is proportional to the coefficient on the sine term in the definition of θ; and the radial extent is proportional to the coefficient on the cosine term in the definition of r. Thus, the curves can be easily tailored to produce the desired effect. Pinwheel-like curves with any number of arms of various shape and length can be created, and sequences of such curves with small increments in the starting phase θ would provide the appearance of whirling pinwheels when animated. We accomplish this by replacing t with $t + s$ in the expression for θ above and then varying s uniformly to produce each successive curve.

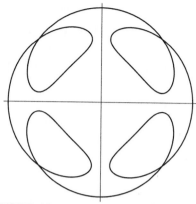

FIGURE 6.5.1. Circular cycloid for $m = n = 4(\phi = 0)$.

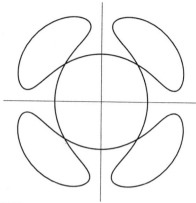

FIGURE 6.5.2. Circular cycloid for $m = n = 4(\phi = \pi)$.

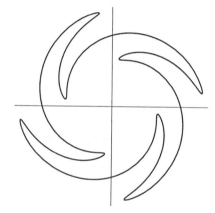

FIGURE 6.5.3. Circular cycloid for $m = n = 4(\phi = \pi/2)$.

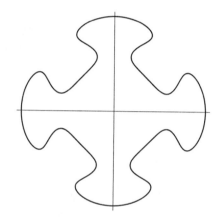

FIGURE 6.5.4. Circular cycloid for $m = 4n = 8(\phi = 0)$.

6.5.3. Curves with Period of r Greater Than Period of θ

If $m < n$ in the definitions of r and θ in Section 6.5.1, then r will have a longer period than θ. For a given m, increasingly higher values of n will introduce more complexity into the curve without altering its fundamental period or number of lobes. We will consider only n that are integer multiples of m. Let us modify the curve in Figure 6.5.1. It has four lobes for $m = 4$, and the loops face inward. Let us change the coefficient c to $1/4$ and increase the argument of the sine term for θ to $8t$ so that $n/m = 2$. The result is shown in Figure 6.5.4.

The lobes can be inverted by adding π to the argument of the cosine term in the definition of r (equivalently changing the sign of the term). The result is shown in Figure 6.5.5; it is simply the rotation of the curve of Figure 6.5.4 by $\pi/4$.

A different curve, shown in Figure 6.5.6, is created if $\pi/2$ is subtracted from the argument of the cosine term in the definition of r (equivalently changing it to a sine term). Adding $\pi/2$ to the argument instead would simply rotate this curve by $\pi/4$.

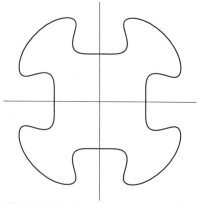

FIGURE 6.5.5.   Circular cycloid for $m = 4$, $n = 8(\phi = \pi)$.

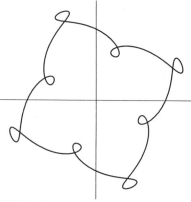

FIGURE 6.5.6.   Circular cycloid for $m = 4$, $n = 8(\phi = -\pi/2)$.

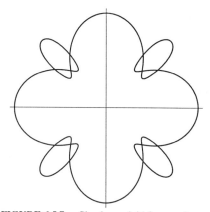

FIGURE 6.5.7.   Circular cycloid for $m = 8$, $n = 4(\phi = 0)$.

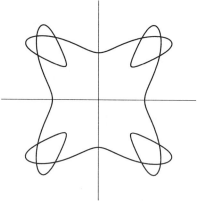

FIGURE 6.5.8.   Circular cycloid for $m = 8$, $n = 4(\phi = \pi)$.

6.5.4.   Curves with Period of r Less Than Period of θ

If $m > n$ in the definitions of r and θ above, then r will have a shorter period than θ. For a given n, increasingly higher values of m will introduce more complexity into the curve without altering its fundamental period or number of lobes. We will consider only m that are integer multiples of n. Again we take the four-lobed curve seen in Figure 6.5.1. Let us change the coefficient c to $1/2$ and increase the argument of the cosine term for r to $8t$ so that $m/n = 2$. The result is given in Figure 6.5.7.

Adding π to the argument of the cosine term for r (or equivalently changing its sign) inverts the curve as seen in Figure 6.5.8. Subtracting just $\pi/2$ from the argument of the cosine term for r (or equivalently changing the cosine to a sine) will give the result in Figure 6.5.9. Adding $\pi/2$ instead would simply rotate this curve by $\pi/4$.

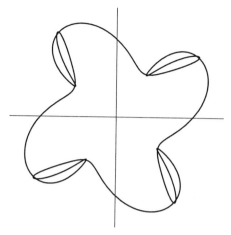

FIGURE 6.5.9. Circular cycloid for $m = 8, n = 4(\phi = -\pi/2)$.

6.6. SUMMARY

In this chapter, we applied periodicity to functions around the circumference of a circle in a manner similar to periodic functions on a line. We found that variations of the simple sine and cosine functions, already used in Chapter 5, produced interesting curves, this time periodic on the circle. Regular polygons were employed as a basis of generating another interesting class of periodic curves on the circle. The trochoid family, analogous to the cycloid family for curves on a line, was presented with numerous examples showing great variation in shape with only small changes in parameters. Lastly, this chapter showed examples of adding harmonic components to the parametric equations for r and θ. Even though the variety of curves generated in this chapter was great, the expressions which generated them contained only a very few terms. Clearly, the possibility exists to create quite intricate and complex periodic curves on the circle if more terms are used.

Chapter 7

SPIRALS

7.1. BASIC SPIRALS

7.1.1. Definition of a Spiral

The previous chapter treated periodic curves on a circle. Such curves closed upon themselves after a given rotation of the angle θ (usually 2π) and would endlessly continue to trace over the connected curve for further rotation of θ. In contrast, spirals will not close on themselves but will continue to drift outward to infinity or inward to zero as the angle θ increases through repeated cycles of 2π. The elementary spirals presented in this chapter can be expressed as

$$r = f(\theta)$$

where $f(\theta)$ is either monotonically increasing or decreasing. However, most of the spirals presented in this chapter can be expressed by the parametric relations

$$r = f(t)$$
$$\theta = g(t)$$

where $f(t)$ is either monotonically increasing or decreasing and $g(t)$ is monotonically increasing. In order to have this property, $f(t)$ cannot be simply a periodic function of t; it requires a factor that is a power of t, a logarithm of t, or an exponential of t. In regard to $g(t)$, the basic specification is $g(t) = t$, meaning that θ is simply linearly increasing; however, $g(t)$ can have additive harmonic components as was done for periodic functions on a circle in the last chapter. We will present many examples in this chapter to illustrate the various spirals that can be generated.

7.1.2. Archimedean Spirals

Named after the famous Greek scientist and engineer, the Archimedean spirals are perhaps the simplest to express. The fundamental *Archimedes' spiral* is given by simply $r = a\theta$. The graph of this curve is formed by a point which moves at a constant rate along a radial line from the origin while the line itself rotates at a constant rate about the origin. If we allow the rate at which the point moves to be variable, there is then a family of Archimedean spirals given by the single relation

$$r = a\theta^m$$

Several spirals in this family have acquired names; see the following table:

m	spiral
1	Archimedes' spiral
$\frac{1}{2}$	Fermat's spiral
-1	hyperbolic spiral
$-\frac{1}{2}$	lituus

FIGURE 7.1.1. Archimedean spirals for $m = 1/2, 1, 2$.

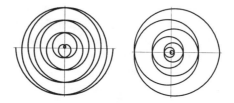

FIGURE 7.1.2. Dual Archimedean spirals for $m = 1, 2$.

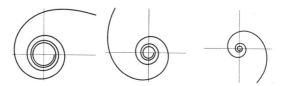

FIGURE 7.1.3. Archimedean spirals for $m = -1/2, -1, -2$.

Let us look first at those spirals for which $m > 0$. Figure 7.1.1 shows plots for $m = 1/2, 1, 2$.

Going outward along any radial line from the origin, the distance from one turn of the spiral to the next is either decreasing for $m < 1$ or increasing for $m > 1$. When $m = 1$ exactly (Archimedes' spiral), the distance between turns is constant.

If negative values of the variable θ are used in addition to positive ones, the plots in Figure 7.1.1 for $m = 1$ and $m = 2$ become symmetric curves as shown in Figure 7.1.2. We will call these curves "dual" Archimedean spirals. In general, when m is odd, the symmetry is about the y axis; and when m is even, the symmetry is about the x axis.

Next we consider the Archimedean spirals for which m is negative. The plots for $m = -1/2, -1, -2$ are shown in Figure 7.1.3. Note that the spirals now converge inward toward the origin with increasing θ. The rate at which this is done is proportional to the exponent m. If negative values of θ are used in addition to the positive ones, the curves for $m = 1$ and $m = 2$ appear as shown in Figure 7.1.4. Just as for positive exponents m, the symmetry is about the y axis when m is odd and about the x axis when m is even.

Any of the Archimedean spirals can be offset with a constant. For instance, take the Fermat spiral ($m = 1/2$) and add a constant to the right-hand side, thus

$$r = a\theta^{1/2} + b$$

This particular form is called the *parabolic spiral*, in recognition of its functional similarity to the parabola, given by the algebraic equation $y = ax^{1/2} + b$. We show

FIGURE 7.1.4. Dual Archimedean spirals for $m = -1, -2$.

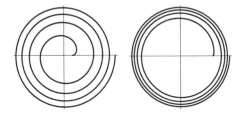

FIGURE 7.1.5. Parabolic spirals for $a = 1, b = 1, 10$.

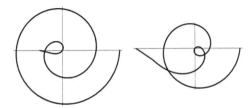

FIGURE 7.1.6. Dual parabolic spirals for $a = 1, b = -1, -2$.

two examples in Figure 7.1.5 for the parabolic spiral with $a = 1$: one with small b and one with large b. Negative values of b will produce curves that must pass through the origin at some value of the variable θ. Such parabolic spirals are lacking in elegance, as seen in the examples in Figure 7.1.6.

7.1.3.   Logarithmic Spiral

The *logarithmic spiral* is given by the equation

$$\ln r = a\theta$$

where ln is the natural logarithm. This equation can be rewritten as

$$r = c\, e^{a\theta}$$

where a constant factor c has been introduced. We see that the value of r increases exponentially with the angle θ. An example of the logarithmic spiral is shown in Figure 7.1.7. Note that because $e^0 = 1$, $r = c$ at $\theta = 0$ in this figure.

The constant a has a meaningful value related to the appearance of the curve. If a radial line from the origin is drawn, then wherever that line intercepts the the curve, the angle α between the curve and the line will be constant. Specifically, $a = \cot \alpha$.

FIGURE 7.1.7. Logarithmic spiral $c = 1, a = 0.1$.

FIGURE 7.1.8. Logarithm spiral variation for $c = 1, a = 0.01, m = 2$.

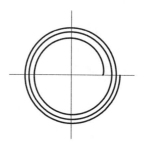

FIGURE 7.1.9. Logarithm spiral variation for $c = 1, a = 0.1, m = 1/2$.

For this reason, the logarithmic spiral is also called the *equiangular spiral*. Another notable property of the logarithmic spiral is the rate at which r increases. If radial lines intercept the curve at (r_1, θ), $(r_2, \theta + \beta)$, and $(r_3, \theta + 2\beta)$, the relation of the lengths, as measured from the origin, of the three radii are

$$r_3/r_2 = r_2/r_1$$

This proportionality holds no matter what the angle increment β or what the exact value is of θ. Another feature of this spiral is the accelerating rate at which r increases. Using the logarithmic expression for this spiral as given above, we can easily write

$$\ln r_2 - \ln r_1 = a(\theta_2 - \theta_1)$$

where r_1 and r_2 are measured at the angles θ_1 and θ_2. Choosing 2π for the difference of the angles (one rotation), the difference of the logarithms of the two radii is simply $2\pi a$. So, if the logarithm of the radius is increased by a constant every full rotation, then the radius itself increases nonlinearly at a greater and greater rate.

A simple modification of the logarithmic spiral would be to use powers of θ, thus

$$r = c \exp(a\theta^m)$$

where *exp* means use the following argument as the exponent on e. Thus, we have a family of logarithmic spirals, just as we had a family of Archimedean spirals. For instance, the curve in Figure 7.1.8 was obtained with $m = 2$ in the above equation. Note that it spirals rapidly outward due to the squaring of the variable θ. In contrast, the rate of spiraling outward is decreased if $m < 1$, as seen in Figure 7.1.9 with $m = 1/2$.

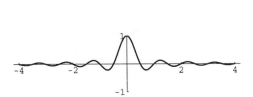

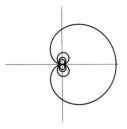

FIGURE 7.1.10. Sinc function with $a = 2\pi$.

FIGURE 7.1.11. Cochleoid.

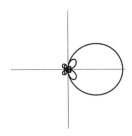

FIGURE 7.1.12. Cochleoid variation for $a = 2$.

FIGURE 7.1.13. Cochleoid variation for $a = 2$ (center part).

7.1.4. Cochleoid

A very simple combination of trigonometric and algebraic functions gives the well-known *sinc function*:

$$y = \sin(ax)/(ax)$$

This function equals unity at $x = 0$ where $\sin(ax)-> ax$ and shows increasingly damped oscillations as x moves away from the y axis (see Figure 7.1.10).

Such a function can be easily adapted to the (r, θ) coordinates with

$$r = \sin(a\theta)/a\theta)$$

When $a = 1$, this generates the beautiful curve called the *cochleoid* shown in Figure 7.1.11. There are two spiral limbs for this case. When a is larger than one, but still integer, the number of spiral limbs will continue to be given by $2a$. For example, Figure 7.1.12 shows the case of $a = 2$. The region of x and y between -0.2 and 0.2 is magnified for clarity in Figure 7.1.13.

7.2. HARMONIC ADDITIONS TO SPIRALS

We would like to produce more character to a spiral by adding some harmonic terms to the r and θ dependence on t in a parametric representation. Taking, for instance, the Archemedean spirals ($r = a\theta^m$) as the base curve, the parametric equations of the enhanced curve will have harmonic terms added to either one or both of the following:

$$r = a(at)^m$$
$$\theta = at$$

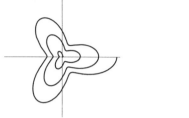
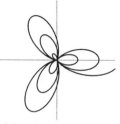

FIGURE 7.2.1. Archimedes' spiral with added harmonic.

FIGURE 7.2.2. Archimedes' spiral with added harmonic.

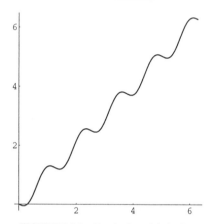

FIGURE 7.2.3. Graph of $t - 0.3\sin(5t)$.

In order to preserve harmony of the additive components with the spiral, the coefficient of the additive component in the radial direction should increase in proportion to the radius of the spiral itself. Thus, the parametric representation of the enhanced spiral will be

$$r = (at)^m [a + b\cos(dt)]$$
$$\theta = at - c\sin(dt)$$

Note that if $b = c = 0$, then $r = a\theta^m$ which is the definition of the Archimedean spirals. Also, we have included the factor d in the sine and cosine arguments; d/a will equal the number of harmonic oscillations per turn of the spiral. An example of the enhanced spiral is given in Figure 7.2.1 for $a = 1, b = 0.5, c = 0.05, d = 3$, and $m = 1$.

For the curve of Figure 7.2.1, $b < a$. If $b = a$, then cusps at the origin are formed at every cycle of the additive harmonic components. This is demonstrated by using the same constants as the curve in Figure 7.2.1, but we change to $b = a = 1$ to produce the curve shown in Figure 7.2.2.

We can vary the factor c in the above definition to produce loops in the harmonic oscillations. This will occur when θ versus t is multivalued along horizontal lines of its graph. For instance, if $a = 1, c = 0.3$, and $d = 5$, the graph of θ versus t appears as in Figure 7.2.3 for the domain of 0 to 2π. Note that, for certain values of θ, t will have three values. Figure 7.2.4 uses $m = 1, a = 1, b = 0.3, c = 0.3$, and $d = 5$ to

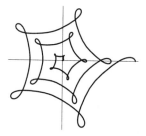

FIGURE 7.2.4. Archimedes' spiral with added harmonic.

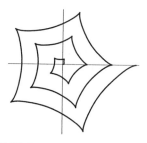

FIGURE 7.2.5. Archimedes' spiral with added harmonic.

FIGURE 7.2.6. Archimedes' spiral with added harmonic.

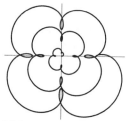

FIGURE 7.2.7. Archimedes' spiral with added harmonic.

demonstrate the appearance of loops. Note that the loops are formed at a rate of d per turn of the spiral.

Is there a relation between c and a for the enhanced spiral at which the loops collapse to cusps? For cusps to form, the derivative of θ with respect to t must equal zero because θ becomes stationary at the exact cusp. From the above definition of θ, the derivative is

$$d\theta/dt = a - cd\cos(dt)$$

Another geometrical feature of the cusps is that they are associated with the extremum of r where $\cos(dt) = 1$. This occurs at $t = (2\pi n)/d$ where $n = 0, 1, 2, \ldots$. The rate at which this occurs is d/a times per turn of the spiral. If $\cos(dt) = 1$ is used in the expression for the derivative above, we have $a - cd = 0$ so that the relation which we sought is $c = a/d$. Figure 7.2.5 shows a plot for exactly this relation using $m = 1$, $a = 1, b = 0.2, c = 0.2$, and $d = 5$; the cusps are indeed present.

It follows that if $c < a/d$, then the cusps will flatten out. In the limit as $c = 0$, the enhanced spiral will show the oscillation of the r component only. If, in addition, $b = 0$, the spiral degenerates to the Archimedean spiral. Figure 7.2.6 shows the case for $c < a/d$ using $m = 1, a = 1, b = 0.2, c = 0.1$, and $d = 5$.

Many modifications of the spiral are possible; and some of the modifications of the basic cycloid in Chapter 5 would serve as a guide here. We will consider only a limited number of them. One useful modification is to let the sign of the cosine term in the r expression be negative in order to turn the loops inward rather than outward. The graph in Figure 7.2.7 illustrates this using $m = 1, a = 1, b = -0.35, c = 0.35$, and $d = 4$.

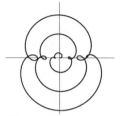
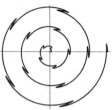

FIGURE 7.2.8. Archimedes' spiral with added harmonic.

FIGURE 7.2.9. Archimedes' spiral with added harmonic.

Another useful modification is to increase or decrease the constants a, b, and c by the same factor while keeping d the same. This has the effect of stretching the pattern of the spiral in the angular direction. Figure 7.2.8 takes the curve parameters from Figure 7.2.7; but the constants a, b, and c are all multiplied by two.

Another interesting modification of the cycloid spiral can be made by letting the coefficient b in the r expression be zero. Then the oscillation is entirely within the θ component; and this gives the spiral a jagged appearance as exemplified by the curve of Figure 7.2.9 which uses $m = 1, a = 2, c = 0.4$, and $d = 10$.

7.3.   COMPLEX SPIRALS

7.3.1.   Spiral of Cornu

A particularly beautiful spiral is based on the integrals of Fresnel, which arise physically in the diffraction of electromagnetic and acoustic waves. This spiral has several names: *spiral of Cornu, Euler's spiral*, and *clothoid*. It is expressed parametrically by

$$x = \int_0^t \sin\left(\frac{\pi\tau^2}{2}\right) d\tau \quad \text{Fresnel's sine integral}$$

$$y = \int_0^t \cos\left(\frac{\pi\tau^2}{2}\right) d\tau \quad \text{Fresnel's cosine integral}$$

The spiral starts at the origin for $t = 0$ and converges on the point $(1/2, 1/2)$ for t large. An antisymmetric copy can be made by simply changing the sign of both integrals. Figure 7.3.1 shows both the positive and negative branches of this spiral.

7.3.2.   Sici Spiral

Another visually striking spiral is the *sici spiral*, as given by the sine and cosine integrals

$$x = \gamma + \ln t + \int_0^t \frac{\cos\tau - 1}{\tau} d\tau \quad \text{cosine integral}$$

$$y = \int_0^t \frac{\sin\tau}{\tau} d\tau \quad \text{sine integral}$$

The sine integral has an asymptotic value of $\pi/2$, and the cosine integral has an asymptotic value of zero. The sici spiral is shown in Figure 7.3.2 with an offset of $-\pi/2$ given to the expression for y in order to make it converge on the origin as t becomes large.

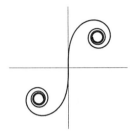

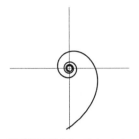

FIGURE 7.3.1. Spiral of Cornu.　　　　　　　　　**FIGURE 7.3.2.** Sici spiral.

7.3.3. Spirals Based on Transcendental Functions—Antisymmetric

When using the parametric-equation approach, the basic "building blocks" for spirals are the sine and cosine functions, which provide the periodicity, and the exponential or power function, which provides the outward growth or the inward decay. Looking at the spiral of Cornu above, it is important to examine whether we can construct similar curves with these basic building blocks because much more flexibility could then be attained by varying constants in these functions. (Moreover, the Fresnel integrals are time-consuming to evaluate.)

Let us try to emulate the effect of the Fresnel integrals with simpler functions. We start by noting that the positive branch of the spiral of Cornu lies in the first quadrant. It converges on the point (1/2, 1/2). The circle given by

$$x = [1 - \cos(at)]/2$$
$$y = [1 + \sin(at)]/2$$

is centered on this point as shown by graph of these equations in Figure 7.3.3. The choice of the start of the circle at $(x, y) = (0, 1/2)$ when $t = 0$, with consequent clockwise motion, will soon be clear. The next step is to incorporate the necessary decay factor:

$$x = [1 - \cos(at)e^{-ct}]/2$$
$$y = [1 + \sin(at)e^{-ct}]/2$$

Plotting these equations with $a = 1$ and $c = 0.1$ now gives the curve in Figure 7.3.4. Just one feature is missing from the curve now—that $y = 0$ at $t = 0$. This can be realized by using the exponential ramp function

$$g(t) = 1 - e^{-bt}$$

as a multiplier of the y function above. This function will force y to zero at $t = 0$ because $e^0 = 1$. Now the parametric equations are

$$x = [1 - \cos(at)e^{-ct}]/2$$
$$y = [1 + \sin(at)e^{-ct}](1 - e^{-bt})/2$$

The result of plotting these equations, and their negatives to produce the antisymmetric branch, is shown in Figure 7.3.5 with $a = 1$, $b = 5$, and $c = 0.1$.

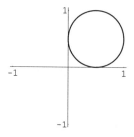

FIGURE 7.3.3. Circle centered on (1/2, 1/2).

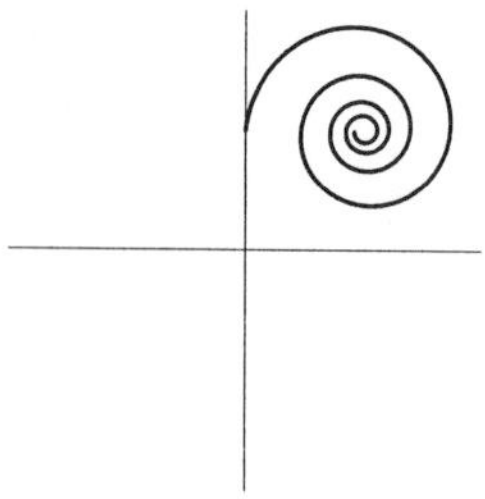

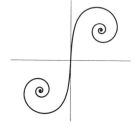

FIGURE 7.3.4. Simple spiral.

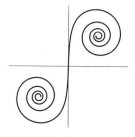

FIGURE 7.3.5. Two-branch spiral.

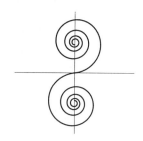

FIGURE 7.3.6. Two-branch spiral.

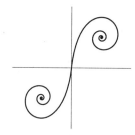

FIGURE 7.3.7. Two-branch spiral.

FIGURE 7.3.8. Two-branch spiral.

We have fairly well achieved the purpose stated above by this last modification. The last set of equations has three constants that can be varied to change the shape of the curve. For instance, a larger value of c should make the spiral arms converge faster to the limit points. Figure 7.3.6 uses the same constants as the one just above, except c is increased from 0.1 to 0.2. By changing b from 5 to 2, a different slant is given to the line passing through the origin (see Figure 7.3.7).

By the same process of building up the desired curve, we can form different spirals. For instance, if we desired to keep the points of convergence on the y axis, then the equations that achieve this are

$$x = \sin(at)e^{-ct}/2$$
$$y = [1 - \cos(at)e^{-ct}]/2$$

Using these equations with $a = 1$ and $c = 0.1$ gives the curve shown in Figure 7.3.8. If we want to move the convergence points toward the origin, we can do so in the

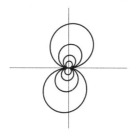
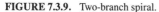

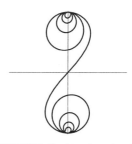

FIGURE 7.3.9. Two-branch spiral.   **FIGURE 7.3.10.** Two-branch spiral.

expression for y by moving the exponential from being a factor on the cosine term to being a factor for the entire expression, thus

$$x = \sin(at)e^{-ct}/2$$
$$y = [1 - \cos(at)]e^{-ct}/2$$

Using these equations with $a = 1$ and $c = 0.1$ gives the curve shown in Figure 7.3.9.

Designing a curve for which the convergence points migrate to $+1$ and -1 on the y axis is somewhat more difficult, but we can achieve this with the following equations:

$$x = \cos(at)e^{-ct}(1 - e^{-bt})/2$$
$$y = \{[\sin(at) - 1]e^{-ct} + 2\}(1 - e^{-bt})/2$$

Using these equations with $a = 1$, $b = 5$, and $c = 0.1$ gives the curve shown in Figure 7.3.10.

These forms are only a few representative spirals that can be generated by combining the exponential and harmonic functions. Many more perturbations of the spiral design are possible. By suitable adjustments of the sign and magnitude of the harmonic terms, appropriate introduction of exponential factors or terms, and use of the exponential ramp, we can meet almost any spiral design criterion. By a process of approximation, we can start with simple curves to satisfy one or two of the design goals and then add more factors or terms to begin to satisfy the other goals. This is not unlike the appoach of the artist who roughly sketches the desired forms and then gradually fills in the details.

7.3.4. Spirals Based on Transcendental Functions—Symmetric
The spirals created in the previous secton were all antisymmetric. It should be noted that all those curves could be converted to symmetric by changing the sign of x only or y only for one of the two branches, depending on whether symmetry about the x or y axis, respectively, was desired. This section treats similar curves, but they are not made symmetric by the simple change of sign just described. The curves presented here do not pass through the origin and can be made antisymmetric only at the cost of losing continuity.

Let us begin with certain design criteria and build a curve in much the same manner as in the previous section. Suppose we want a spiral that will have one branch decay

FIGURE 7.3.11. Simple spiral.   **FIGURE 7.3.12.** Two-branch spiral.

counterclockwise toward $(0, 1/2)$ and the other decay clockwise toward $(0, -1/2)$. Start with the simple spiral which converges to $(0, 1/2)$:

$$x = [\cos(at)e^{-ct}]/2$$
$$y = [1 + \sin(at)e^{-ct}]/2$$

The plot of this function is shown in Figure 7.3.11 for $a = 1$ and $c = 0.1$. It is clear that by simply reversing the sign of the y expression, the lower branch can be formed from the same equations. The only missing part is the connecting line between the starting points of the two branches. This is created by using the exponential ramp as a factor in the y expression, thus

$$x = [\cos(at)e^{-ct}]/2$$
$$y = [1 + \sin(at)e^{-ct}](1 - e^{-bt})/2$$

By plotting these equations with $a = 1$, $b = 5$, $c = 0.1$ and repeating with the sign of the y expression reversed, we achieve the desired curve shown in Figure 7.3.12.

By moving the exponential decay factor in the expression for y so that it is a factor on the whole expression, we can make the spirals converge toward the origin from both sides. The equations are then

$$x = [\cos(at)e^{-ct}]/2$$
$$y = [1 + \sin(at)](1 - e^{-bt})e^{-dt}/2$$

The graph of these parametric functions is shown in Figure 7.3.13 using the same constants as the previous plot and $d = 0.1$. (To get a pleasing visual effect, the constants c and d should be equal.)

Moving the convergence points outward to $(0,1)$ and $(0,-1)$ is somewhat more difficult to analyze; however, the following equations do this:

$$x = \cos(at)e^{-ct}/2$$
$$y = \{[\sin(at) - 1]e^{-ct} + 2\}(1 - e^{-bt})/2$$

Using these equations with $a = 1$, $b = 5$, and $c = 0.1$ gives the curve shown in Figure 7.3.14.

The curves shown so far have all had $x = 1/2$ at $y = 0$. If we wanted to move the connection between the upper and lower branches outward along the x axis, another

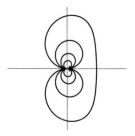

FIGURE 7.3.13. Two-branch spiral.

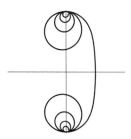

FIGURE 7.3.14. Two-branch spiral.

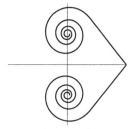

FIGURE 7.3.15. Two-branch spiral.

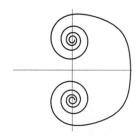

FIGURE 7.3.16. Two-branch spiral.

term is needed in the x expression. We choose an exponential term that is large at $t = 0$ but decays rapidly. In this way none of the former curve for moderate to large values of t is altered appreciably. The equations are then

$$x = [\cos(at)e^{-ct} + e^{-dt}]/2$$
$$y = [1 + \sin(at)e^{-ct}](1 - e^{-bt})/2$$

Figure 7.3.15 shows the curve formed by these equations with $a = 1, b = 5, c = 0.1$, and $d = 5$.

The behavior for small t is linear for both x and y causing the undesirable kink at (1, 0). This behavior could be predicted by examining the limit of the functions for small t. In order to get a smooth appearance, we let the exponent of the second term for the x component be $-dt^2$ rather than simply $-dt$. The curve, with $a = 1, b = 5, c = 0.1$, and $d = 5$, now matches the one in Figure 7.3.16.

What if we want the direction of the spirals to be reversed while maintaining the connecting line between the two branches on the positive side of the x axis? Returning to the first spiral formed in this section, we achieve this by simply changing the sign of the sine term in the y expression and using the ramp factor, thus

$$x = [\cos(at)e^{-ct}]/2$$
$$y = [1 - \sin(at)e^{-ct}](1 - e^{-bt})/2$$

Figure 7.3.17 shows the plot of this parametric function.

As for the antisymmetric spirals in the previous section, many varieties of symmetric spirals can be generated by suitable usage of the harmonic and exponential

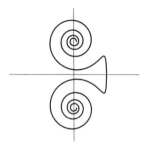

FIGURE 7.3.17. Two-branch spiral.

functions. The key to achieving a design goal is, again, to build up the expression of the curve through successive approximations to the goal.

7.4. SUMMARY

Spirals are among the most esthetically pleasing curves and have many realizations in the natural world. This chapter has introduced the simplest spirals first. By adding harmonic components to the parametric expressions for simple spirals, we can generate many more interesting and useful spiral shapes with cusps, loops, and periodic lobes on them. The last part of this chapter showed how we can create complex symmetric and antisymmetric spirals of many shapes by the careful use of harmonic and exponential functions. The design possibilities are only limited by our imagination and our time constraints.

Chapter 8

NONPERIODIC FUNCTIONS ON A LINE

8.1.   DEFINITION

Chapter 5 presented curves that were periodic on the line x. Simply stated, these curves repeated with a basic period L such that $y(x) = y(x + L)$ for any x. We shall describe as *nonperiodic* any curves that do not have this property on the domain of $-\infty < x < +\infty$. Chapters 2, 3, and 4 have already treated the elementary curves of this broad class: polynomials, radicals, and exponentials. The purpose of this chapter is to look at more complicated forms which satisfy the nonperiodic pattern on the line x. These forms will entail a combination of harmonic, exponential, or polynomial parts. We have already learned all that is necessary to construct the curves of this chapter, but the explicit treatment here of many of these nonperiodic curves should enable us to more easily and readily design such curves for particular needs.

8.2.   HARMONIC CURVES WITH NONLINEAR ARGUMENTS

8.2.1.   Case of Sine Curves

If we generalize the harmonic sine curve to

$$y = \sin[aP(x)]$$

where $P(x)$ is a simple monomial of x, written as x^m, then the harmonic function will no longer be periodic, except for the case of $m = 1$. The effect will be to progressively stretch or compress the strictly harmonic curve as x increases in either the positive or negative direction from the origin. The curves for powers of x from one through four are illustrated in Figures 8.2.1 to 8.2.4, respectively. Note that each has a different x scale to better present the behavior of each curve.

These curves are symmetric for even powers of x and antisymmetric for odd powers of x. The value of the sine function at $x = 0$ is always zero in these cases. For powers of x greater than one, the derivative will also be zero at $x = 0$. On the basis of these properties, note that we can always change the symmetry by plotting the curve in the negative domain of x with the opposite sign. The result will remain continuous and smooth at $x = 0$. (The original harmonic with power of x equal to one is an exception—it would be continuous at $x = 0$ but not smooth.)

If we use fractional powers of x less than one, the effect will be to stretch the curves at large x. Figures 8.2.5 and 8.2.6 show the harmonic function with powers of x equal to 3/4 and 1/2, respectively. For negative x, we have used the absolute value of x to calculate the root. These curves are continuous at $x = 0$ but not smooth. Note that we can make these curves antisymmetric by plotting the negative of the function over the negative x domain. The resulting curves will be both continuous and smooth at $x = 0$.

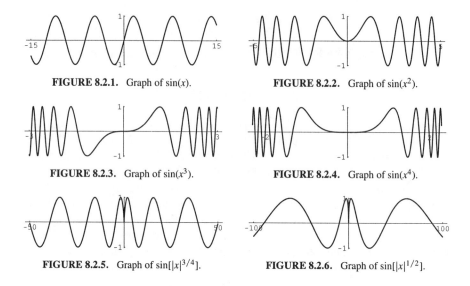

FIGURE 8.2.1. Graph of $\sin(x)$.   **FIGURE 8.2.2.** Graph of $\sin(x^2)$.

FIGURE 8.2.3. Graph of $\sin(x^3)$.   **FIGURE 8.2.4.** Graph of $\sin(x^4)$.

FIGURE 8.2.5. Graph of $\sin[|x|^{3/4}]$.   **FIGURE 8.2.6.** Graph of $\sin[|x|^{1/2}]$.

8.2.2. Case of Cosine Curves

Similar to the sine curve, we can generalize the harmonic cosine curve to

$$y = \cos[aP(x)]$$

where $P(x)$ is a simple monomial of x, written as x^m. This function will no longer be periodic, except for the case of $m = 1$. Figures 8.2.7 to 8.2.10 illustrate this function for powers of x equal to one through four, respectively. Again, we use varying x scales to show the curves better.

The cosine curves are all symmetric about the y axis and cannot be made antisymmetric without producing a discontinuity at $x = 0$. (Including other factors in the function, such as the exponential ramp of Chapter 4, would enable us to do this but not without significantly changing the curve's shape near $x = 0$.) If we use fractional powers of x less than one, the cosine wave at large x then produces a progressively more and more stretched wave. Figures 8.2.11 and 8.2.12 illustrate this stretching using powers equal to 3/4 and 1/2, respectively. The absolute value of x at negative x is used to enable us to calculate the roots.

Although it may not be clear at the scale of these plots, these cosine-like functions are smooth at $x = 0$ for fractional powers of x. This is because the derivative is given by a sine function, which is zero at $x = 0$. As for the cosine-like curves with powers of x greater than one, we cannot make these fractional-power curves antisymmetric without losing continuity.

8.3. SINC FUNCTION AND ITS VARIATIONS

One of most well-known functions in applied mathematics is the *sinc* function, introduced in Section 7.1.4 and having the form $\sin(x)/x$. This function is interesting in that it apparently has a singularity at $x = 0$. However, in looking at the series

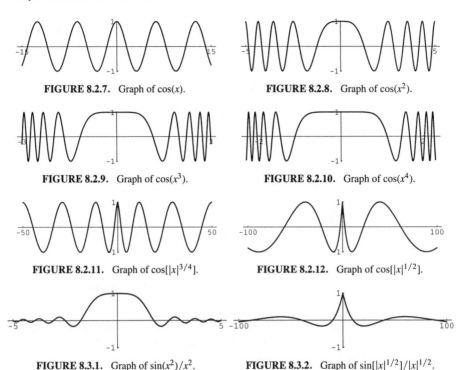

FIGURE 8.2.7. Graph of $\cos(x)$.

FIGURE 8.2.8. Graph of $\cos(x^2)$.

FIGURE 8.2.9. Graph of $\cos(x^3)$.

FIGURE 8.2.10. Graph of $\cos(x^4)$.

FIGURE 8.2.11. Graph of $\cos[|x|^{3/4}]$.

FIGURE 8.2.12. Graph of $\cos[|x|^{1/2}]$.

FIGURE 8.3.1. Graph of $\sin(x^2)/x^2$.

FIGURE 8.3.2. Graph of $\sin[|x|^{1/2}]/|x|^{1/2}$.

expansion of the sine function,

$$\sin(x) = x - x^3/3! + x^5/5! - x^7/7! + \cdots,$$

one divides each term by x and finds that the limit of the sinc function at $x = 0$ is in fact unity. As long as the argument of the sine function is equal to the denominator, we can use different powers of x and still expect the function to equal unity at $x = 0$. Figure 8.3.1 shows the graph when x^2 is substituted for x. In Figure 8.3.2, we use the power equal to 1/2. In order to be able to calculate the function for $x < 0$, we use the absolute value of x rather than x itself.

A parametric form involving the sinc function leads to some interesting curve shapes. Consider

$$x = at + b\sin(t)$$
$$y = \sin(t)/t$$

This curve may have multiple values on the y axis. They occur wherever $x = 0$ or

$$at + b\sin(t) = 0$$

Taking the expression for x and letting $c = -b/a$, the y values at $x = 0$ would be given by

$$t = c\sin(t)$$

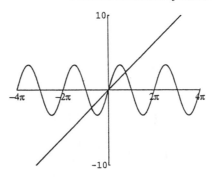

FIGURE 8.3.3. Solutions of $t = (10/3)\sin(t)$.

FIGURE 8.3.4. Sinc function variant.   **FIGURE 8.3.5.** Sinc function variant.

Clearly, $t = 0$ is a solution. Other solutions are attained only numerically. We can give a graphical representation of the solutions by plotting the function t and the function $c \cdot \sin(t)$ and seeing where they cross. For instance, if $a = 1/10$ and $b = -1/3$, the plots would look like those in Figure 8.3.3. Solutions occur for this case at roughly $-3\pi/4$ and $+3\pi/4$. If we substitute these values back into the expression for y above, the y values will be identical and equal to roughly 0.3.

As the coefficient c on the sine function increases, the straight line in the graph of Figure 8.3.3 will cross more peaks and troughs of the sine wave and add four solutions each time. Figure 8.3.4 shows the parametric sinc variant with the coefficients $a = 1/10$ and $b = -1/3$. Note that there are the two identical crossings at roughly $y = 0.3$, as predicted.

Using positive b requires only a small change in the analysis for the y-axis crossings. For positive b, the sine wave in Figure 8.3.3 will change sign. For the same exact coefficients, the straight line will fail to intercept the sine wave at all, leaving the solution at $t = 0$ as the only solution. Figure 8.3.5 shows the parametric sinc variant for $a = 1/10$ and $b = 1/3$.

8.4. HARMONIC CURVES COMBINED WITH EXPONENTIALS

From the known behavior of the exponential function (Chapter 4), it should be relatively clear how ordinary, or more complicated, harmonics will look when we apply an exponential factor to them. The simplest case would be the cosine function multiplied by $\exp(-|x|)$ shown in Figure 8.4.1. We will look at some more complex examples of such functions involving harmonics and exponentials here. What if you basically wanted an oscillating function translated away from the x axis but then smoothly

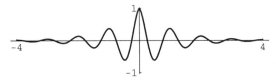

FIGURE 8.4.1. Cosine wave decay with application of $\exp(-|x|)$ factor.

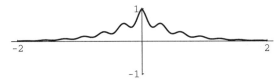

FIGURE 8.4.2. Upward translated cosine wave decay with $\exp(-2|x|)$.

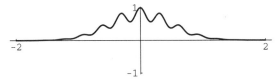

FIGURE 8.4.3. Upward translated cosine wave decay with $\exp(-2x^2)$.

decaying with increasing x? Let the translated sine wave be given by

$$y = a + b\cos(cx)$$

In order to make this curve decay, we employ the simple exponential function of Chapter 4 to get

$$y = [a + b\cos(cx)]e^{d|x|}$$

where d is understood to be a negative number. Figure 8.4.2 shows an example using the coefficients $a = 0.8, b = 0.2, c = 20$, and $d = -2$. This curve is continuous at $x = 0$ but not smooth. We can achieve smoothness by using the exponential function with a squared argument:

$$y = [a + b\cos(cx)]\exp(dx^2)$$

where again the coefficient d is understood to be negative. Figure 8.4.3 illustrates this function for $a = 0.5, b = 0.2, c = 20$, and $d = -2$.

What if you basically wanted a sine function translated away from the x axis but converging on the origin regardless of the amount of translation? Let the translated sine wave be given by

$$y = a\,\text{sign}(x) + b\sin(cx)$$

where $\text{sign}(x)$ means -1 for $x < 0$ and $+1$ for $x > 0$. As defined, this function has a discontinuity of $2a$ units at the origin. In order to make this curve pass through the origin, we employ the exponential ramp of Chapter 4 to get

$$y = [a\,\text{sign}(x) + b\sin(cx)](1 - e^{d|x|})$$

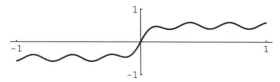

FIGURE 8.4.4. Graph of $[0.5 \, \text{sign}(x) + 0.1 \, \sin(20x)][1 - \exp(-15|x|)]$.

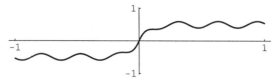

FIGURE 8.4.5. Graph of $[0.5 \, \text{sign}(x) + 0.1 \, \text{sign}(x) \cos(20x)][1 - \exp(-15|x|)]$.

where d is understood to be a negative number. Figure 8.4.4 shows an example using the coefficients $a = 0.5, b = 0.1, c = 20$, and $d = -15$. Figure 8.4.5 shows the same concept, but it is now applied to the cosine wave instead with parameters identical to the curve of Figure 8.4.4. For this plot, we also apply the $\text{sign}(x)$ factor to the coefficient b.

We can see how the exponential function can be used to ramp functions up or down thereby achieving different shapes. The coefficient in the exponential term controls the rate or slope of the ramp function and can be varied until the desired effect is achieved.

8.5. VARIATIONS ON CYCLOIDS

8.5.1. Cycloids with Exponentials

The cycloids of Chapter 5 were a rich source of periodic curve shapes. Here, we will look again at the cycloids, only now with exponential or monomial weighting, which makes them no longer periodic. First, we reiterate the general parametric equations of the cycloid

$$x = at - b\sin(t)$$
$$y = b\cos(t)$$

One of the obvious modifications to make in this regard is to weight each of the harmonic terms with an exponential, thus

$$x = at - b\sin(t)e^{c|t|}$$
$$y = d\cos(t)e^{c|t|}$$

where the coefficient c is understood to be negative. Note that we have also allowed the coefficients on the two harmonic terms to be different. Figure 8.5.1 illustrates this modified cycloid for $a = 0.05, b = -0.2$, and $c = -0.1$, and $d = 1.0$. Changing only the sign of the coefficient b produces a different effect, as seen in Figure 8.5.2.

It may be desirable to have different constants in the exponential function for the x and y components. The plot in Figure 8.5.3 keeps the width of the loops nearly

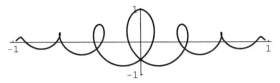

FIGURE 8.5.1. Cycloid with exponential weighting.

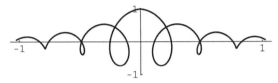

FIGURE 8.5.2. Cycloid with exponential weighting.

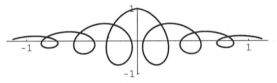

FIGURE 8.5.3. Cycloid with exponential weighting.

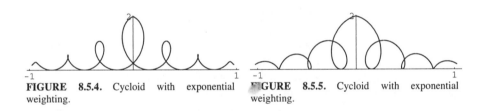

FIGURE 8.5.4. Cycloid with exponential weighting.

FIGURE 8.5.5. Cycloid with exponential weighting.

uniform by having a much smaller constant for the x component ($c = -0.02$) than for the y component ($c = -0.1$). The remaining parameters are the same as for the plot in Figure 8.5.2.

By offsetting the y function and letting the exponential multiply the entire expression, we get

$$x = at - b\sin(t)e^{c|t|}$$
$$y = d[1 + \cos(t)]e^{c|t|}$$

The effect of this latest modification is to make the mimima of the cycloid always lie on the x axis. Figure 8.5.4 and 8.5.5 show examples with the same coefficients used for the previous plots: first with negative b and then with positive b.

The examples of this section have a discontinuity of slope at $x = 0$ due to using $|t|$ as the argument of the exponential function. Using t^2 as the argument instead would eliminate this discontinuity and make the curves smooth at $x = 0$.

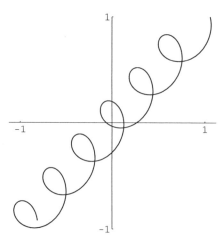

FIGURE 8.5.6. Cycloid with a linear *y* term.

8.5.2. Cycloids with Monomial Terms in *Y*

Again taking the basic cycloid given at the beginning of the previous section, we can incorporate additional *t* dependence in the *y* expression with

$$x = at - b\sin(t)$$
$$y = ct^n + b\cos(t)$$

For instance, if $n = 1$, there will be a linear dependence, as illustrated in Figure 8.5.6 for $a = 0.05$, $b = 0.2$, and $c = 0.05$. Even though this curve is not strictly periodic in *x*, it is clear that it is periodic along the line $x = y$ and can be rotated to produce a curve which is periodic in *x*. When $n = 2$ (quadratic term), the curve becomes symmetric about the *y* axis. Figure 8.5.7 illustrates this with $a = 0.05$, $b = 0.2$, and $c = 0.0025$. (A change of sign for *b* in the *y* expression was also made here.)

Any value of *n* can be used in the above. Even values of the integer *n* will produce curves symmetric about the *y* axis while odd values will produce curves with no symmetry at all. For odd *n*, symmetry can be achieved by taking the absolute value of *t* in the first term of the *y* component, as shown for $n = 1$ in Figure 8.5.8.

8.6. CURVES WITH HARMONICS IN *Y* AND POLYNOMIALS IN *X*

An interesting family of parametric curves is formed when we use polynomials for the *x* expression and a harmonic for the *y* expression. Here are the general parametric equations:

$$x = P(t)$$
$$y = \sin(bt + \phi)$$

The function $P(t)$ will be a polynomial of arbitrary degree given by the following form:

$$P(t) = a_1 t + a_2 t^2 + a_3 t^3 + \cdots$$

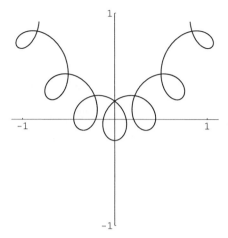

FIGURE 8.5.7. Cycloid with a quadratic y term.

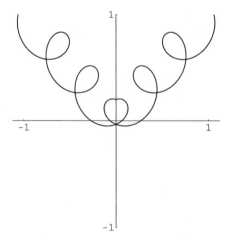

FIGURE 8.5.8. Cycloid with absolute value of a linear y term.

Note that no constant term is included. The case of all a_i equal zero except a_1 will produce the ordinary sine wave and is therefore trivial. The case of all a_i equal zero except one (not the first) will produce those curves already treated in the first section of this chapter. Therefore, we are only interested in those cases where at least two of the coefficients are nonzero.

Chapter 2 showed that polynomial curves are symmetric only when the nonzero coefficients have even-numbered indices in the expression $P(t)$ and that they are antisymmetric only when the non-zero coefficients have odd-numbered indices. Here, we will consider only antisymmetric forms for $P(t)$ to make the parametric curve either symmetric or antisymmetric. Thus, our selection of polynomials is now limited to

$$P(t) = a_1 t + a_3 t^3 + a_5 t^5 + \cdots$$

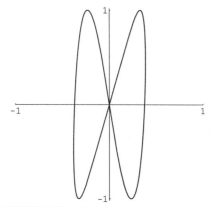

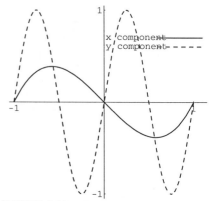

FIGURE 8.6.1. Parametric curve—polynomial in *x* and harmonic in *y*.

FIGURE 8.6.2. Components of the parametric curve.

This, when combined with $\phi = 0$ in the general equations above, will produce anti-symmetric curves and, combined with $\phi = \pi$, will produce symmetric curves.

Chapter 2 also revealed that, in the limit of large *t*, the polynomials $P(t)$ will behave as the largest power of *t*. This means that, at some moderate value of *t*, the polynomial will effectively behave as a monomial and we would merely be looking at the curve forms treated in the first section of this chapter. Therefore, we will plot our current equations for small to moderate values of *t* where the interesting behavior can be seen.

As an example, take $a_1 = -1$ and $a_3 = 1$ along with $b = 2\pi$ and $\phi = 0$ in the above general expressions. The curve appears as shown in Figure 8.6.1 for $-1 < t < 1$. Now that we have plotted it, can we explain what we see? To do so, we must look at each component of the parametric equation. Figure 8.6.2 shows these components as *t* varies between -1 and 1. How the curve in Figure 8.6.1 is generated can now be seen from the behavior of the two functions. This curve can be stretched or compressed along the *x* axis by applying a constant factor to the function $-t + t^3$. It can be translated along the *x* axis by adding a constant to $-t + t^3$ and changing the phase of *y* accordingly.

To really modify the curve, we must change the coefficients of $-t + t^3$ in a non-proportional manner. Let us try $-t + 2t^3$; the result is shown in Figure 8.6.3 for $-1 < t < 1$. Values of a_3 between 1 and 2 will place the endpoints on the *x* axis between the limit of zero for $a_3 = 1$ and unity for $a_3 = 2$. Again, note that the parametric curves made with the cubic function for *x* and a sine function for *y* are antisymmetric.

Let us now change from a sine function to a cosine function for *y*. Again, we start with equal coefficients to get $x = -t + t^3$ and produce the curve in Figure 8.6.4 for $-1 < t < 1$. This curve also crosses over itself, but now at $y = 1$. In order to really change the appearance of the curve, we use unequal coefficients. The result for $a_1 = 1$ and $a_3 = 2$ is shown in Figure 8.6.5.

For a given polynomial, it is difficult to predict the range of *x* for a given range of *t*. There is a class of polynomials, called *Legendre polynomials*, which are rather well-behaved in the sense that their range is -1 to 1 for the domain of -1 to 1. They

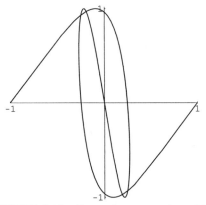

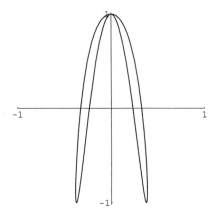

FIGURE 8.6.3. Parametric curve—polynomial in *x* and harmonic in *y*.

FIGURE 8.6.4. Parametric curve—polynomial in *x* and harmonic in *y*.

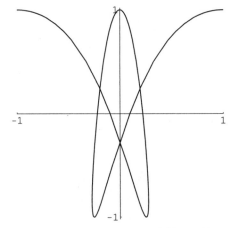

FIGURE 8.6.5. Parametric curve—polynomial in *x* and harmonic in *y*.

involve either all even powers of *x* or all odd powers of *x*. The first five Legendre polynomials are

$$P_1 = t$$
$$P_2 = (-1 + 3t^2)/2$$
$$P_3 = (-3t + 5t^3)/2$$
$$P_4 = (3 - 30t^2 + 35t^4)/8$$
$$P_5 = (15t - 70t^3 + 63t^5)/8$$

The monomial relation is of no interest here. Also, the even-order polynomials cannot be used in the present context for the reason already mentioned. Using the third Legendre polynomial for *x* along with the cosine function for *y*, we get the curve shown in Figure 8.6.6 for $-1 < t < 1$. Using the fifth Legendre polynomial for *x* and the cosine function for *y*, we get the curve shown in Figure 8.6.7 for $-1 < t < 1$.

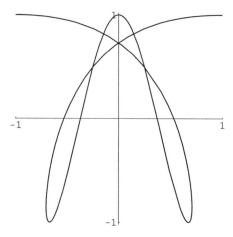

FIGURE 8.6.6. Parametric curve—third Legendre polynomial in x and harmonic in y.

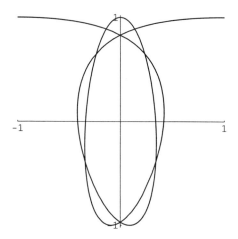

FIGURE 8.6.7. Parametric curve—fifth Legendre polynomial in x and harmonic in y.

8.7. CURVES WITH POLYNOMIALS IN BOTH X AND Y

If we use polynomials for both x and y expressions in the parametric curve form, we can write this as

$$x = f(t)$$
$$y = g(t)$$

where f and g are both polynomials. The range of possibilities with this form is limitless as the polynomials can be taken to high degree. However, here we will only look at some low-degree polynomials. In particular, the Legendre polynomials introduced in the previous section have the nice property of being limited to a range of -1 to 1 for the same domain; and they consequently provide a good basis for showing the possibilities.

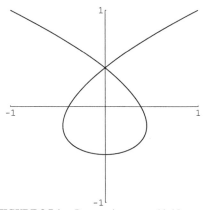

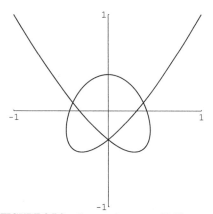

FIGURE 8.7.1. Parametric curve—third Legendre polynomial in x and second Legendre polynomial in y.

FIGURE 8.7.2. Parametric curve—third Legendre polynomial in x and fourth Legendre polynomial in y.

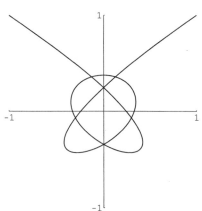

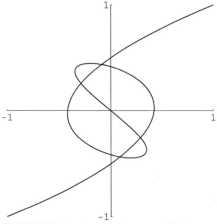

FIGURE 8.7.3. Parametric curve—fifth Legendre polynomial in x and fourth Legendre polynomial in y.

FIGURE 8.7.4. Parametric curve—fifth Legendre polynomial in x and third Legendre polynomial in y.

For x, the odd-degree polynomials must be used to attain symmetry or antisymmetry in the curves. In the case of y, the odd-degree polynomials will then give a antisymmetric curve while the even-degree polynomials will give a symmetric curve. Taking the simplest case, we plot $y = P_2(t)$ against $x = P_3(t)$ in Figure 8.7.1. Figure 8.7.2 shows $y = P_4(t)$ against $x = P_3(t)$, and Figure 8.7.3 shows $y = P_4(t)$ against $x = P_5(t)$. The last example, shown in Figure 8.7.4, is of $y = P_3(t)$ against $x = P_5(t)$ and is antisymmetric, in contrast to the preceding symmetric examples.

There is no reason to limit ourself to the Legendre polynomials for designing these types of curves. Also, changing the constants in these polynomials will produce variations of these curves. Another class of useful polynomials with nearly the same properties as the Legendre ones are the Chebyshev polynomials, but they will not be illustrated here.

8.8. SUMMARY

This chapter has illustrated many functions that are nonperiodic on the x axis but which are more complex than basic curves in previous chapters. We showed that the harmonic functions themselves can be modified to generate nonperiodic curves by changing the arguments to be nonlinear. The sinc function and several variations of it were presented. The use of exponentials with harmonic functions was illustrated with many examples to show the effect of damping due to the exponential. We subjected the cycloid and its variations to the same type of damping to produce new and interesting shapes. We showed how polynomials in x, rather than exponential functions, can be used to vary the shape of periodic curves and make them nonperiodic. Lastly, we introduced some curves based on polynomials for both the x and y components; and we showed that certain well-known polynomials such as the Legendre polynomials can be used effectively here.

Chapter 9

CLOSED CURVES WITHOUT PERIODICITY

9.1. DEFINITION

Chapter 6 treated many curves that could be plotted around the circle; their common property was that they were all periodic in θ: the angle measured counterclockwise from the positive x axis. All the examples shown there also had the property of being closed curves, although examples of curves that are periodic on the circle but not closed can be found. A curve $f(\theta)$ is *closed* if its endpoints, $f(a)$ and $f(b)$, coincide when it is plotted over the interval $a \leq \theta \leq b$. For example, the circle is a closed curve while a spiral is not. The many closed curves that are not periodic are the subject of this chapter. By periodic, we mean more than two lobes because many of the curves in this chapter will have two lobes and are thus, in a strict sense, periodic with period π.

Closed curves may either be simple or complex. Simple closed curves have no multiple points (except the endpoints). The circle is an elementary example of a simple closed curve. The equation of the circle of radius one can be written as

$$y = (1 - x^2)^{1/2}$$

provided that we plot both the positive and negative roots. We can give this simple closed curve multiple points by merely multiplying the equation with x:

$$y = x(1 - x^2)^{1/2}$$

The graph of this function is shown in Figure 9.1.1 next to the original circle.

This chapter will treat both simple and complex closed curves. Most will be symmetrical about either the x or y axis. As stated for the curves of Chapter 8, if you have studied Chapters 2 through 4, you already have acquired the knowledge to construct the curves to be presented. But, by explicitly showing many examples, we hope to show the range of possibilities from basic smooth forms to complex, intricate curves.

9.2. CIRCLES AND ELLIPSES WEIGHTED WITH HARMONICS

The equation for an ellipse can be solved for y:

$$y = b[1 - x^2/a^2]^{1/2}$$

where a and b are the *semi-axes* of the ellipse (b/a gives the height over the width). In the following we will always let $b = 1$. The ellipse is a simple closed curve with two zero crossings of the x axis: at $-a$ and $+a$. Can the ellipse be twisted about the x axis to be a complex closed curve in some simple manner? Obviously, the effect of multiplying by x, as shown in Figure 9.1.1 for the circle, will be similar for an ellipse. To achieve more zero crossings, recall that the sine and cosine functions are

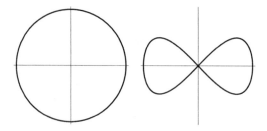

FIGURE 9.1.1. Examples of simple and complex closed curves.

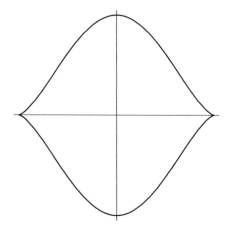

FIGURE 9.2.1. Graph of $\cos(\pi x/2)\sqrt{(1 - x^2)}$.

periodic with predictable zero crossings. By using them as multiplicative factors on the above, we can introduce the desired twisting of any order. Let the equation now be expressed as

$$y = b\cos(2\pi cx)[1 - x^2/a^2]^{1/2}$$

or

$$y = b\sin(2\pi cx)[1 - x^2/a^2]^{1/2}$$

The constant c will determine the number and position of the new zero crossings.

First consider the cosine factor. Because the ellipse extends to $+a$ on the positive x axis, the first zero crossing of the cosine function should come at no greater than $+a$. If the cosine function is to equal zero at $x = a$, then

$$\cos(2\pi ca) = 0$$

which requires $2\pi ca = \pi/2$. Thus, $c = 1/(4a)$ in order for the first zero of the cosine function to fall at the edge of the ellipse; and so we will require $c > 1/(4a)$ to introduce new zero crossings between 0 and $+a$. If the cosine factor is always required to be zero at $x = a$, then one sets

$$c = (2n - 1)/(4a) \quad (n = 2, 3, 4, \ldots)$$

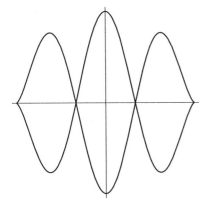
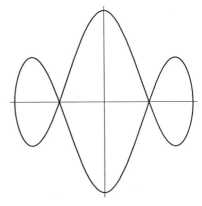

FIGURE 9.2.2. Graph of $\cos(3\pi x/2)\sqrt{(1-x^2)}$.   **FIGURE 9.2.3.** Graph of $\cos(\pi x)\sqrt{(1-x^2)}$.

to produce $n-1$ additional zero crossings on the positive x axis. Due to the symmetry of the cosine function, all this will have the same effect on the negative x side of the ellipse. To begin with, look at the limiting case in Figure 9.2.1 where $a = b = 1$ and $c = 1/4$ exactly. To get one intermediate zero crossing for $a = b = 1$, we let $c = 3/4$; Figure 9.2.2 shows the graph for this.

Values of c between the two just used will show interesting effects. The special values

$$c = n/(2a) \quad (n = 1, 2, 3, ...)$$

will create curves that are smooth everywhere. Look at the graph in Figure 9.2.3 for $c = 1/2$ and $a = b = 1$. Notice that the curve appears to be smooth at the limits of $-a$ and $+a$ on the x axis. We can, in fact, show that the derivative is continuous at these points.

The analysis for sine weighting of the ellipse is very similar to that for the cosine weighting. Recall that the sine function is zero at $x = 0$; there will always be at least one new zero crossing introduced when a sine factor is brought in. If the sine factor is required to be zero for the edge of the ellipse at $x = a$, then

$$\sin(2\pi ca) = 0$$

which requires $2\pi ca = \pi$. Thus, $c = 1/(2a)$ in order for the first zero of the sine function to fall at the edge of the ellipse; and so we will require $c > 1/(2a)$ in order to introduce new zero crossings in addition to the one at $x = 0$. If the sine factor is always required to be zero at $x = a$, then one sets

$$c = n/(2a) \quad (n = 2, 3, 4, ...)$$

to produce $n-1$ additional zero crossings on the positive x axis. Due to the antisymmetry of the sine function, all of this pertains to the negative x axis also. Let's first look at the limiting case where $c = 1/(2a)$ with $a = b = 1$; this produces the expected zero crossing at $x = 0$ only, as seen in Figure 9.2.4.

To introduce a zero crossing on the x axis between 0 and 1, we let $c = 1/a$ with $a = b = 1$. Figure 9.2.5 shows the graph. As for the cosine case, intermediate values

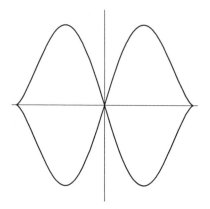

FIGURE 9.2.4. Graph of $\sin(\pi x)\sqrt{(1-x^2)}$.

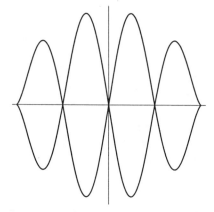

FIGURE 9.2.5. Graph of $\sin(2\pi x)\sqrt{(1-x^2)}$.

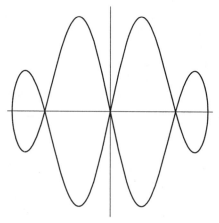

FIGURE 9.2.6. Graph of $\sin(3\pi x/2)\sqrt{(1-x^2)}$.

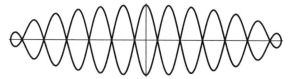

FIGURE 9.2.7. Graph of $\cos(3\pi x/2)\sqrt{(1-x^2/4^2)}$.

of c will produce interesting effects. The curve shown in Figure 9.2.6 is for $c = 3/(4a)$ with $a = b = 1$. Note the smoothness of the function at the limits of $x = -a$ and $x = +a$; we can show that the derivative here is continuous.

As a final example in this section, we construct many zero crossings in the cosine-weighted ellipse by using $c = 3/4$ with $a = 4$ and $b = 1$; this gives $n = 6$ zero crossings between 0 and $x = 4$ as seen in Figure 9.2.7.

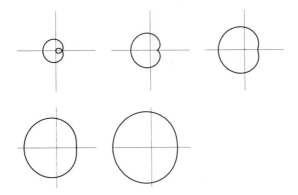

FIGURE 9.3.1. Hypotrochoids for $a = 1$, $b = 2$, $c = 1, 2, 3, 4, 5$.

9.3. TROCHOIDS WITHOUT PERIODICITY

9.3.1. Hypotrochoids

Chapter 6 presented a family of curves called trochoids. Although the trochoids normally constructed are periodic around the circle, there is a suite of parameters for which the trochoids are not periodic in appearance. You are reminded that "periodic" in this context means that there is some angle α, for which 2π is an even multiple, through which the curve can be rotated about the origin and exactly overlap the original curve. We restate the equations of the hypotrochoid:

$$x = (a - b)\cos(t) + c\cos[(a - b)t/b]$$
$$y = (a - b)\sin(t) - c\sin[(a - b)t/b]$$

Detailed description of the hypotrochoid can be found in Section 6.4.1. Given that a/b is reduced to its lowest terms, all cases treated in that section involved $a > 1$ because the number of lobes was equal to a and the curve would not be periodic, in the sense of the current definition, for $a = 1$. Now let us look at those cases where $a = 1$.

To begin with the simplest possibility, the plots of Figure 9.3.1 show $a = 1$ and $b = 2$ for increasing values of c. Note the loop in the first of the plots in Figure 9.3.1 and the fact that a cusp occurs when $c = b$. The next group of plots in Figure 9.3.2 represent $a = 1$ and $b = 3$. Again, the cusp occurs where $c = b$. But there is now a loop within a loop for $c = 1$; this gradually deforms into a single loop for large c.

The next set of plots in Figure 9.3.3 shows $a = 1$ and $b = 4$. Again, there is the cusp at $c = b$. The loops for $c = 1$ are nested three deep and, for c large, they are nested two deep. The latter remains true regardless of how large c becomes. The pattern that has emerged is that for $c < b$ loops are nested $b - 1$ deep and that for $c > b$ they are nested $b - 2$ deep when $a = 1$. At the intermediate value of $c = b$, a cusp will appear on the innermost loop.

9.3.2. Epitrochoids

The epitrochoid was introduced in Section 6.4.2. A full description of the epitrochoid is given there; however, we restate its equations here:

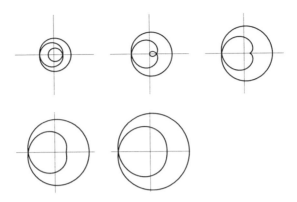

FIGURE 9.3.2. Hypotrochoids for $a = 1$, $b = 3$, $c = 1, 2, 3, 4, 5$.

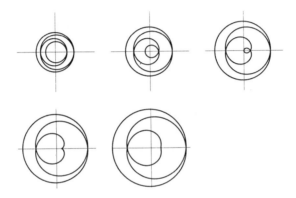

FIGURE 9.3.3. Hypotrochoids for $a = 1$, $b = 4$, $c = 1, 2, 3, 4, 5$.

$$x = (a + b)\cos(t) - c\cos[(a + b)t/b]$$
$$y = (a + b)\sin(t) - c\sin[(a + b)t/b]$$

As for the hypothrochoid, the cases having no periodicity for the epitrochoid require that $a = 1$ when a/b is reduced to its lowest terms. We will examine these cases here.

First taking $a = 1$ and $b = 2$, the curve can take the forms shown in Figure 9.3.4. Compared with the case of $a = 1$ and $b = 2$ for the hypotrochoid, these plots show the opposite behavior as c is increased. The plot for $c < b$ gives one loop while for the plot for $c > b$ gives a loop within a loop. But just as for the $a = 1$ and $b = 2$ case for the hypotrochoid, the cusp appears when $c = b$.

Let us next look at the epitrochoids for $a = 1$ and $b = 3$ plotted in Figure 9.3.5. These epitrochoids for $a = 1$ and $b = 3$ can be compared to the hypotrochoids for $a = 1$ and $b = 3$. Again note that an opposite dependence on c appears but that the cusp remains at $c = b$. The pattern that emerges for the epitrochoids of $a = 1$ is that there will be a nesting of loops $b - 1$ deep for $c < b$ and that this nesting becomes b deep for $c > b$. In between, at $c = b$, a cusp will appear.

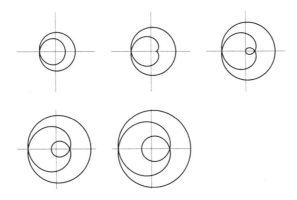

FIGURE 9.3.4. Epitrochoids for $a = 1$, $b = 2$, $c = 1, 2, 3, 4, 5$.

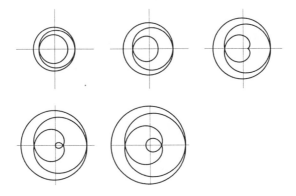

FIGURE 9.3.5. Epitrochoids for $a = 1$, $b = 3$, $c = 1, 2, 3, 4, 5$.

9.4. LISSAJOUS CURVES

The *Lissajous curves* (also called *Bowditch curves*) comprise one of the most striking and varied families of curves having the closed property. They are generated with the seemingly simple parametric relations

$$x = \sin[(a/c)2\pi t + b\pi]$$
$$y = \sin(2\pi t)$$

where, to satisfy the closed property, a/c is rational and b is integer. To make a complete curve, it is required that x be identical at $t = 0$ and at some $t = \tau$ (an integer). Furthermore, if the curve is to be smooth at these endpoints, the derivative of x must be equal at $t = 0$ and $t = \tau$. At these endpoints, there is also the requirement that both y values be identical and equal to zero and that both derivatives of y be identical and equal to unity. The requirement that x be identical at the endpoints is written as

$$\sin[(a/c)2\pi\tau + b\pi] = \sin(b\pi)$$

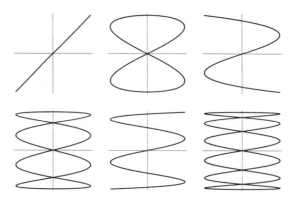

FIGURE 9.4.1. Lissajous curves for $c = 1$, $b = 0$, $a = 1, 2, 3, 4, 5, 6$.

or, expanding the left-hand side,

$$\sin[(a/c)2\pi\tau]\cos(b\pi) + \cos[(a/c)2\pi\tau]\sin(b\pi) = sin(b\pi)$$

Because b is assumed to be an integer, $\cos(b\pi) = \pm 1$ and $\sin(b\pi) = 0$; the above equation then reduces to

$$\sin[(a/c)2\pi\tau] = 0$$

The choice of $\tau = c$ is clear. At this value, the requirements for both x and y will be met; thus, to make a complete curve, $0 < t < c$. We assume in making this choice that a/c has been reduced to its lowest terms.

We will show many examples of the Lissajous curves to illustrate their great variety. At the end of this section, the results will be summarized in a concise manner. Let us begin with $b = 0$ and $c = 1$ and show the variation with parameter a. For $b = 0$, the Lissajous equations reduce to

$$x = \sin[(a/c)2\pi t]$$
$$y = \sin(2\pi t)$$

These Lissajous curves are shown in Figure 9.4.1. Note that for a an odd integer, the curves appear to collapse to a single line. This line is antisymmetric about the x axis. If we used $b = 1$ in the phase term, the curves for odd a would reflect about the y axis because the sign of the x expression would be reversed. For even a, there is both x and y symmetry, and a change to $b = 1$ would not alter the curves. An interesting case is $b = 1/2$, for which the Lissajous equations reduce to

$$x = \cos[(a/c)2\pi t]$$
$$y = \sin(2\pi t)$$

The curves in Figure 9.4.2 keep the same parameters as those in Figure 9.4.1 except b is changed to $1/2$. In contrast to $b = 0$, the curves which collapse to a line occur for even a. Also, in constrast to $b = 0$, the $b = 1/2$ curves are symmetric about the x axis

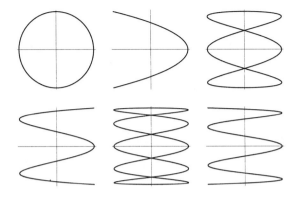

FIGURE 9.4.2. Lissajous curves for $c = 1$, $b = 1/2$, $a = 1, 2, 3, 4, 5, 6$.

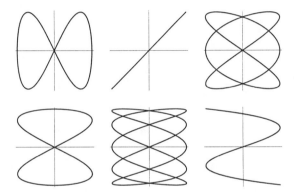

FIGURE 9.4.3. Lissajous curves for $c = 2$, $b = 0$, $a = 1, 2, 3, 4, 5, 6$.

rather than antisymmetric. The curves for odd a have the same symmetry properties as those for even a when $b = 0$. If b were changed from 1/2 to 3/2, the sign of the x expression would be reversed. This would reflect the curves for even a about the y axis but leave the curves for odd a unchanged.

The next series of curves for $c = 2$ and $b = 0$ are shown in Figure 9.4.3. Those curves that collapse to lines are for $a/c = 1$ and 3, the odd integers. These are also the ones that would be flipped about the y axis if b were changed to 1.

The next series in Figure 9.4.4 uses the same parameters, but with $b = 1/2$. Now the curve that collapses to a line is the one for which $a/c = 2$, the only even integer among the six cases. This curve too is the only one affected by a change of b from 0 to 1.

We continue in Figure 9.4.5, but now with $c = 3$ and $b = 0$. The curves that collapse to a line are those for which a/c is a ratio of odd integers. The next suite of curves in Figure 9.4.6 uses the same parameters except b is changed to 1/2. Now the curves that collapse are those whose ratio a/c is an even integer over an odd integer.

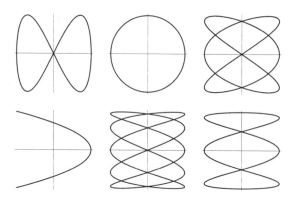

FIGURE 9.4.4. Lissajous curves for $c = 2$, $b = 1/2$, $a = 1$, 2, 3, 4, 5, 6.

FIGURE 9.4.5. Lissajous curves for $c = 3$, $b = 0$, $a = 1$, 2, 3, 4, 5, 6.

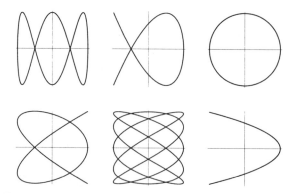

FIGURE 9.4.6. Lissajous curves for $c = 3$, $b = 1/2$, $a = 1$, 2, 3, 4, 5, 6.

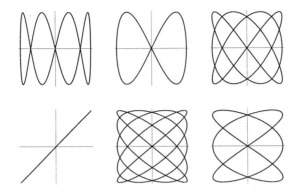

FIGURE 9.4.7. Lissajous curves for $c = 4$, $b = 0$, $a = 1, 2, 3, 4, 5, 6$.

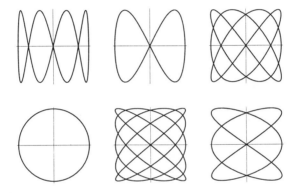

FIGURE 9.4.8. Lissajous curves for $c = 4$, $b = 1/2$, $a = 1, 2, 3, 4, 5, 6$.

Next we look at curves for which $c = 4$ and $b = 0$ in Figure 9.4.7. Lastly, we look at the curves for the same parameters as Figure 9.4.7, but with $b = 1/2$ as plotted in Figure 9.4.8. By increasing c, we can make the Lissajous curves more and more complex.

We have now seen sufficient examples to generalize the behavior of Lissajous curves whose phase is 0, $\pi/2$, π, or $3\pi/2$. We find that:

1. The arbitrary fraction a/c will produce the same curve as the same fraction reduced to its lowest terms. For instance, $a/c = 6/3$ gives a curve identical to $a/c = 2/1$. Any following statements assume a/c has been so reduced.
2. The phase factor $b = 0$ produces the same graph as $b = 1$ for given a and c except for reflection about the y axis. Depending on the symmetry, this reflection may, in fact, result in the same curve appearance. This statement also is valid for the pair $b = 1/2$ and $b = 3/2$.
3. For $b = 0$ (or 1):

    a) For a odd and c odd:
    - all curves are collapsed to a line

- $a > c$ gives a curve antisymmetric about the x axis
- $a < c$ gives a curve antisymmetric about the y axis
- $a = c = 1$ gives a straight line midway between the axes

b) For a even and c odd:

- all curves are symmetric about both axes
- all curves have c nodes (crossing points of the curve) on the x axis
- all curves have $a - 1$ nodes on the y axis plus intercepts on the y axis at ± 1

c) For a odd and c even:

- all curves are symmetric about both axes
- all curves have a nodes on the y axis
- all curves have $c - 1$ nodes on the x axis plus intercepts on the x axis at ± 1

d) Any curve for a given a/c is identical to one for c/a, except for a rotation by π and, in the case of a and c both odd, reflection about an axis.

4. For $b = 1/2$ (or $3/2$):

a) For a odd and c odd:

- all curves are symmetric about both axes
- all curves have $a - 1$ nodes on the y axis plus intercepts on the y axis at ± 1
- all curves have $c - 1$ nodes on the x axis plus intercepts on the x axis at ± 1
- $a = c = 1$ is a circle

b) For a even and c odd:

- all curves are collapsed to a line
- all curves are symmetric about the x axis
- all curves have a intercepts on the y axis
- all curves have $(c - 1)/2$ nodes on the x axis plus one intercept on the x axis at -1 or $+1$

c) For a odd and c even:

- all curves are symmetric about both axes
- all curves have a nodes on the y axis
- all curves have $c - 1$ nodes on the x axis plus intercepts on the x axis at ± 1

Thus far, only b factors that give the phases 0, $\pi/2$, π, and $3\pi/2$ have been used. For other, intermediate b values, the curves have different shapes. Let us look at the curves for $b = 1/4$. For this value, the Lissajous equations reduce to

$$x = \{\sin[(a/c)2\pi t] + \cos[(a/c)2\pi t]\}/\sqrt{2}$$
$$y = \sin(2\pi t)$$

Figures 9.4.9 to 9.4.12 show suites of curves that use $b = 1/4$ and have $c = 1, 2, 3, 4$, respectively. The complexity, symmetry, and other behavior of these curves is predicable. An examination of these curves, along with those previously shown, indicates that a Lissajous curve will cross the y axis $2a$ times and the x axis $2c$ times; some crossings may be at identical points which is consistent with less complex curves. Therefore, in general, the complexity will increase as the integers in the ratio a/c increase. To illustrate this principle, we plot a suite of curves for $a/c = 5/7$ with

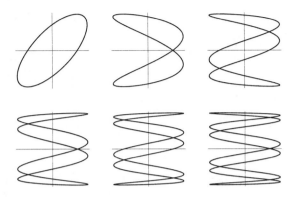

FIGURE 9.4.9. Lissajous curves for $c = 1$, $b = 1/4$, $a = 1, 2, 3, 4, 5, 6$.

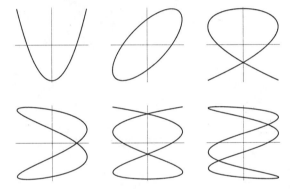

FIGURE 9.4.10. Lissajous curves for $c = 2$, $b = 1/4$, $a = 1, 2, 3, 4, 5, 6$.

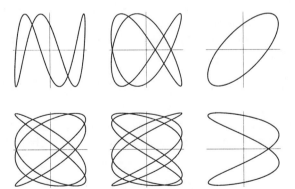

FIGURE 9.4.11. Lissajous curves for $c = 3$, $b = 1/4$, $a = 1, 2, 3, 4, 5, 6$.

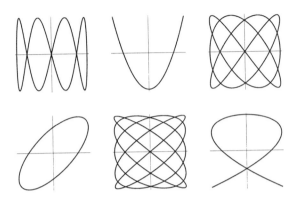

FIGURE 9.4.12. Lissajous curves for $c = 4$, $b = 1/4$, $a = 1, 2, 3, 4, 5, 6$.

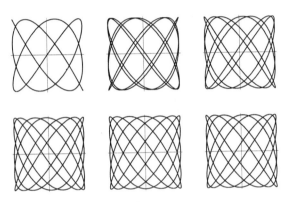

FIGURE 9.4.13. Lissajous curves for $a/c = 5/7$, $b = 0, 1/8, 1/4, 3/8, 1/2, 5/8$.

$b = 0$ to 5/8 in increments of 1/8 in Figure 9.4.13. Note that the $b = 5/8$ curve is a reflection of the $b = 3/8$ curve about the y axis; such symmetry would also occur for the $b = 3/4$ and 1/4 pair and the $b = 7/8$ and 1/8 pair. Shifting to $a/c = 6/7$, the suite of curves in Figure 9.4.14 shows that the value of $b = 0$ does not necessarily impose any less complexity on the plotted curve.

9.5.   VARIATIONS ON LISSAJOUS CURVES

9.5.1.   *Y* Given by Powers of Sine Wave

In this section, variations on the Lissajous curves are examined. These variations will be implemented by modifying the y expression only in the basic Lissajous curve given by

$$x = \sin[(a/c)2\pi t + b\pi]$$
$$y = \sin(2\pi t)$$

The first variation will be to allow y to be a power of the sine wave: $\sin^2$, $\sin^3$, ... Note that when the power is odd, the sine term will still vary between -1 and $+1$;

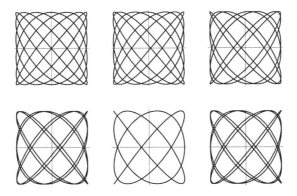

FIGURE 9.4.14. Lissajous curves for $a/c = 6/7$, $b = 0$, $1/8$, $1/4$, $3/8$, $1/2$, $5/8$.

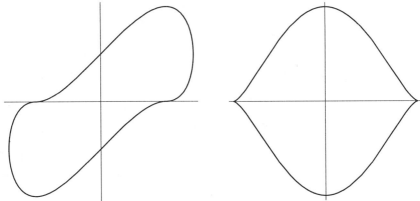

FIGURE 9.5.1. Lissajous-like curve for $a/c = 1/1$, $b = 1/4$, $n = 2$.

FIGURE 9.5.2. Lissajous-like curve for $a/c = 1/1$, $b = 1/2$, $n = 3$.

however, when it is even, the sine term will always be positive. For the even powers, we can still achieve the negative and positive values by multiplying the result by the sign of the value of the sine itself at the same argument. Thus, for an even integer, we use

$$y = \sin^n(2\pi t) \, \text{sign}[\sin(2\pi t)]$$

For $n = 2$, $a/c = 1/1$, and $b = 1/4$, the curve in Figure 9.5.1 is produced. Figure 9.5.2 shows an example using $n = 3$, $a/c = 1/1$, and $b = 1/2$. A more complex example, shown in Figure 9.5.3, has $n = 3$, $a/c = 2/3$, and $b = 1/4$.

The shapes that can be created with this variation are far too diverse to show more than a few examples here. The result, compared to the ordinary Lissajous curve with identical parameters, will usually be less graceful.

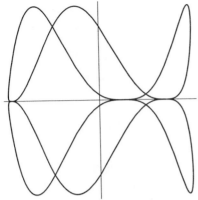

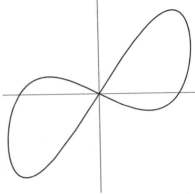

FIGURE 9.5.3. Lissajous-like curve for $a/c =$ 2/3, $b = 1/4, n = 3$.

FIGURE 9.5.4. Lissajous-like curve for $a/c =$ 1/1, $b = 0, n = 2$.

9.5.2. *Y* Given by Sum of Harmonics

The next variation of the Lissajous curve uses a harmonic in the *y* term; thus the *y* component is

$$y = [\sin(2\pi t) + \sin(n2\pi t + d\pi)]/2$$

where *n* is an integer greater than or equal to two. For the sake of simplification, we will only give examples with $d = 0$ here. For example, we use $n = 2$, $a/c = 1/1$, and $b = 0$ in Figure 9.5.4. Recall that these same parameters in the equations for the ordinary Lissajous curve produced a straight line cutting diagonally between the two axes.

The example in Figure 9.5.5 uses the same parameters used in Figure 9.5.4 except *b* is changed to 1/4. In contrast to all ordinary Lissajous curves, this curve possesses no symmetry. The example in Figure 9.5.6 uses the same parameters as in Figure 9.5.5 except *a* is changed to 2. The last example using an additional harmonic term in *y* has $a/c = 1/1$, $b = 1/2$, and $n = 3$ as seen in Figure 9.5.7.

Even with these few examples, we can see that this variation of the Lissajous curve can lead to complex and intricate curves, which may resemble freehand drawing in some cases. The prediction of the curve shape will be difficult, indeed, and will require some experimentation with the parameters to achieve a desired shape.

9.5.3. *Y* Given by Product of Harmonics

If *y* is given as a product of harmonics

$$y = \sin(2\pi t) \sin(n2\pi t + d\pi)$$

then the *y* expression of the ordinary Lissajous curve is modulated. For the sake of simplicity, we will consider only examples where $d = 1/2$, making the modulating factor a cosine factor. Figure 9.5.8 shows the result when $n = 2$. The same parameters, but with $n = 3$, produces the graph seen in Figure 9.5.9. Making $b = 1/2$ and using the same parameters in Figure 9.5.9 produces the graph seen in Figure 9.5.10.

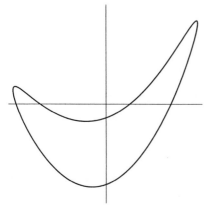

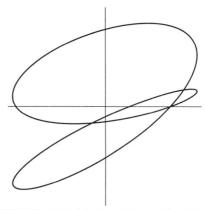

FIGURE 9.5.5. Lissajous-like curve for $a/c = 1/1, b = 1/4, n = 2$.

FIGURE 9.5.6. Lissajous-like curve for $a/c = 2/1, b = 1/4, n = 2$.

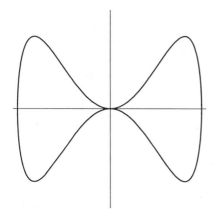

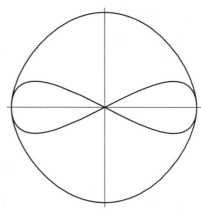

FIGURE 9.5.7. Lissajous-like curve for $a/c = 1/1, b = 1/2, n = 3$.

FIGURE 9.5.8. Lissajous-like curve for $a/c = 2/1, b = 0, n = 2$.

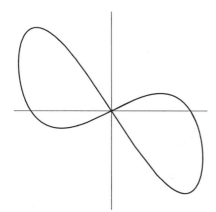

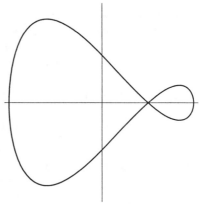

FIGURE 9.5.9. Lissajous-like curve for $a/c = 2/1, b = 0, n = 3$.

FIGURE 9.5.10. Lissajous-like curve for $a/c = 2/1, b = 1/2, n = 3$.

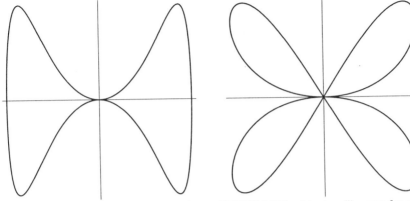

FIGURE 9.5.11. Lissajous-like curve for $a/c = 1/1$, $b = 1/2$.

FIGURE 9.5.12. Lissajous-like curve for $a/c = 2/1$, $b = 0$.

As for the cases where y is given by a sum of harmonics, this variation of the Lissajous curve can produce many interesting shapes. But, again, the prediction of the curve shape is not straightforward and would require some experimentation to achieve a desired shape.

9.5.4. *Y Given by Compound Harmonic*

The last variation of Lissajous curves we will examine involves the compound harmonic. Let y be given as

$$y = \sin[n\pi \sin(2\pi t)]$$

For the sake of simplicity, we will consider only the case where $n = 1$. For the first example, let $a/c = 1/1$ and $b = 1/2$; the result is shown in Figure 9.5.11. Recall that the ordinary Lissajous curve with these same parameters was the circle. The next example in Figure 9.5.12 uses $a/c = 2/1$ and $b = 0$. This is more complex than the ordinary Lissajous curve with the same parameters. Figure 9.5.13 shows another example for $a/c = 2/1$ and $b = 1/4$.

Again, the variation of the Lissajous equation for y will produce interesting shapes, but they are difficult to predict. For this particular modification, as well as all the previous ones, it may be beneficial to plot the x and y expressions separately. We can then adjust the parameters while looking at the individual x and y plots to help predict the final shape of the Lissajous-like curve.

9.6. ADDITIONAL PARAMETRIC CURVES WITH SYMMETRY

In this subsection, we will examine a more complex set of parametric equations given by

$$x = a\sin(2\pi t + k\pi) + b\sin(p\pi t + l\pi)$$
$$y = c\sin(2\pi t + m\pi) + d\sin(q\pi t + n\pi)$$

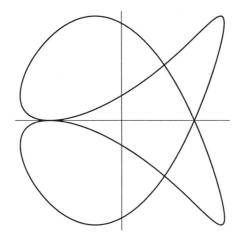

FIGURE 9.5.13. Lissajous-like curve for $a/c = 2/1$, $b = 1/4$.

where the coefficients k, l, m, and n are taken to be either 0 or 1/2 thereby implying that the terms will be ordinary sines or cosines. The factors p and q are restricted to integers. These equations enable one to construct a large variety of shapes with one or, in special cases, two axes of symmetry. Because Chapter 6 dealt with forms having more than one axis of symmetry, the focus here is shapes having only one. Recall that the basic ellipse can be expressed parametrically as

$$x = a\cos(2\pi t)$$
$$y = b\sin(2\pi t)$$

The curves generated with the more complex parametric forms above will, in a sense, be deformations of the basic ellipse. Once this is perceived, it is easy to design a curve shape of interest.

For instance, suppose we wanted an egg-shaped curve. Starting with an ellipse, we just need to stretch it out at one end while compacting it at the other. This can be simulated by adding a cosine term to the y expression but with twice the frequency:

$$x = a\cos(2\pi t) + b\cos(4\pi t)$$
$$y = c\sin(2\pi t)$$

The choice of $a = 1.0$, $b = 0.1$, and $c = 0.7$ produces a shape closely resembling an egg in Figure 9.6.1. A larger value of b will make one end become concave, as illustrated by the shape in Figure 9.6.2, where $a = 1.0$, $b = 0.5$, and $c = 0.7$ are used. Stretching out the y values and compressing the x values will produce a different effect. Figure 9.6.3 uses $a = 0.2$, $b = 0.3$, and $c = 1.0$.

These curves all used $d = 0$. If we allow d to be nonzero, then the range of possible shapes broadens even more. The parametric equations to be used are now

$$x = a\cos(2\pi t) + b\cos(4\pi t)$$
$$y = c\sin(2\pi t) + d\sin(4\pi t)$$

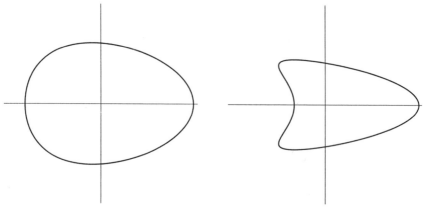

FIGURE 9.6.1. Egg-shaped curve.   **FIGURE 9.6.2.** Arrowhead-shaped curve.

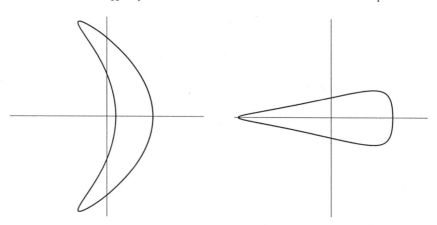

FIGURE 9.6.3. Boomerang-shaped curve.   **FIGURE 9.6.4.** Carrot-shaped curve.

An example using $a = 1.0$, $b = -0.2$, $c = 0.3$, and $d = 0.1$ is shown in Figure 9.6.4.

In order to see how the final shape will appear, it is beneficial to plot the x and y expressions separately. For the curve in Figure 9.6.4, these would give the two graphs shown in Figure 9.6.5. By adjusting the x and y curves separately, we can quickly approximate the coefficients required to produce a desired shape from the parametric equations.

So far, we have purposely avoided any shapes with loops. However, the parametric equations will allow this to happen with the proper choice of coefficients. For example, if d is increased to 0.3 in the curve of Figure 9.6.4, then a curve with a loop, shown in Figure 9.6.6, is the result. Is it possible to predict this behavior? Let us plot the separate components of this curve shown in Figure 9.6.7. Note that in this case, the expression for y crosses the t axis four times rather than twice. These additional zero crossings produce the loop seen in the curve of Figure 9.6.6. Additional loops can be formed by increasing the frequency of the second term for y.

Thus far, only harmonics whose frequency was a multiple of the fundamental were

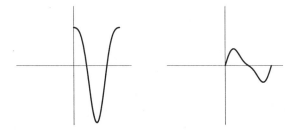

FIGURE 9.6.5. Parametric components of carrot-shaped curve.

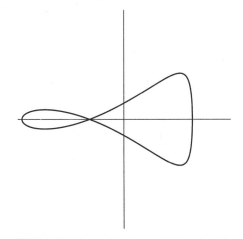

FIGURE 9.6.6. Carrot-shaped curve deformed to a loop.

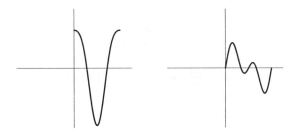

FIGURE 9.6.7. Parametric components of curve with loop.

used in the *x* or *y* expressions; this implies that *p* or *q* is an even integer. Certainly, a choice of an odd integer is acceptable; however, this will require that the range of *t* be extended to [0, 2] rather than [0, 1] to produce a complete curve. The use of an odd integer can produce a whole new variety of shapes, usually more complex than those with even integers. For instance, the graph in Figure 9.6.8 uses the same parameters as the "boomerang" curve in Figure 9.6.3, but now *p* is changed from 4 to 3.

Another modification of the basic ellipse is to weight one or the other of the parametric components with another harmonic expression. Consider the following

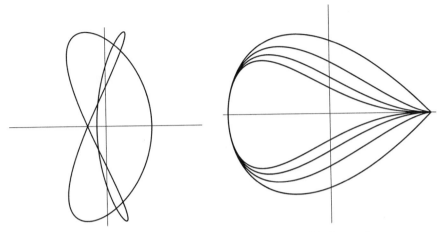

FIGURE 9.6.8. Parametric curve.   **FIGURE 9.6.9.** Parametric curves.

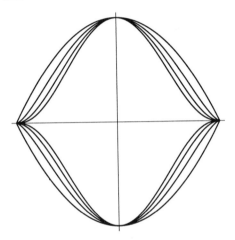

FIGURE 9.6.10. Parametric curves.

equations:

$$x = a\cos(2\pi t)$$
$$y = b\sin(2\pi t)|\sin(m\pi t)|^n$$

For $m = 1$, the weighting will be zero at $t = 0$ and $t = 1$ so that one side of the ellipse will be squeezed down to the x axis. For $a = b = 1$, the graphs of Figure 9.6.9 show the effect of $n = 1$, 2, 3, and 4 together. If we change to $m = 2$, then both sides of the ellipse will be squeezed. Figure 9.6.10 shows the same cases as Figure 9.6.9 but now for $m = 2$. An arbitrary value of m, assumed to be integer, will force the basic ellipse to the x axis at m points.

Recall the trochoid family of Section 6.4. These functions on the circle had the

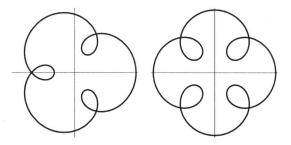

FIGURE 9.6.11. Hypotrochoid for $k = 4, 5$.

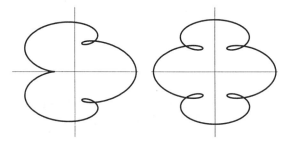

FIGURE 9.6.12. Hypotrochoid-like curve for $k = 4, 5$.

property of periodicity. We write a special form of the hypotrochoid simply as

$$x = \cos(t) + a \cdot \cos(kt)$$
$$y = \sin(t) + a \sin(kt)$$

where k is an integer, giving $k - 1$ lobes for the hypotrochoid. For instance, the figures generated with $k = 4$ and $k = 5$ are shown in Figure 9.6.11 for $a = 1/2$.

These hypotrochoids are naturally periodic on the circle. What happens if we instead use

$$x = \cos(t) + a \cos(kt)$$
$$y = \sin(t) + b \sin(kt)$$

where b does not equal a? The result with $a = 1/2$ again but $b = 1/4$ is shown in Figure 9.6.12. These curves are now symmetric but not periodic as the true hypotrochoids.

Many more symmetric curves can be generated using variations of the examples given in this section. They are all closely related to the trochoids introduced in Chapter 6 and share some of the simple beauty of that family of curves.

9.7. CLOSED CURVES USING HARMONICS BUT LACKING SYMMETRY

Often, it may be desirable to produce a closed curve that has no symmetry at all. One of the simplest ways to accomplish this is to add a harmonic to either of the

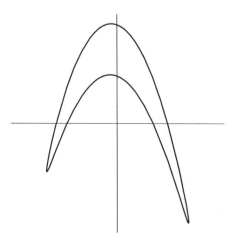

FIGURE 9.7.1. Parametric asymmetric curve.

parametric components of the ellipse when it is rotated by an angle $p\pi$. A curve was already generated in Section 9.5.2 with equations relating to the Lissajous curves. The harmonic starts and ends with zero amplitude at $t = 0$ and $t = 1$, respectively, and also has zero derivative at both endpoints. The equations to be used are

$$x = a\cos(2\pi t + p\pi)$$
$$y = c\sin(2\pi t) + d\sin(m\pi t)$$

We should realize that the choice of m is not arbitrary. First, it must be an integer in order for the curve to be closed. Second, this integer must be even to avoid having a discontinuity in the derivative at the endpoints $t = 0$ and $t = 1$. Furthermore, only those even integers that are an even multiple of 2 will produce the asymmetric shape. This is because when $m = 6, 10, 14, \ldots$, the y value at t is equal to the negative of that at $t + 1/2$; thus, the ellipse is skewed by an equal and opposite amount at all t. Formally, by expanding $y(t + 1/2)$, we have

$$y(t + 1/2) = -c\sin(2\pi t) + d\sin(m\pi t)\cos(m\pi/2)$$

If one of $m = 6, 10, 14, \ldots$ is used in this equation, then the cosine factor is -1 and the sign of d will be changed to negative. On the other hand, if $m = 4, 8, 12, \ldots$, the sign of d remains positive. Only in the later case will the curve be asymmetric. Figure 9.7.1 shows the asymmetric curve when $a = 1$, $p = 1/4$, $c = 1/2$, $d = 1$, and $m = 4$. Figure 9.7.2 shows another example using $a = 1$, $p = 1/8$, $c = 1/2$, $d = 0.3$, and $m = 8$.

We can switch to a cosine term to be added to the y expression

$$x = a\cos(2\pi t + p\pi)$$
$$y = c\sin(2\pi t) + d\cos(m\pi t)$$

This produces different effects on the basic ellipse. We can show that the same rules apply regarding the integer m. Figure 9.7.3 shows an example using $a = 1$, $p = 1/4$, $c = 1.0$, $d = 0.25$, and $m = 4$.

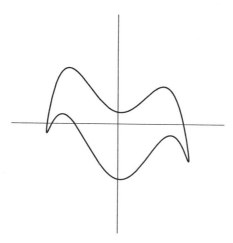

FIGURE 9.7.2. Parametric asymmetric curve.

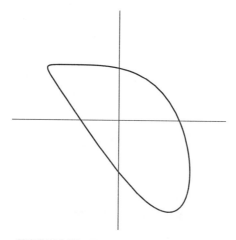

FIGURE 9.7.3. Parametric asymmetric curve.

These are only a few examples of closed curves having no symmetry. By applying the functions of previous chapters, there are numerous other ways to achieve a lack of symmetry. The nice property about the approach with harmonics here is that the curves possess a continuous first derivative everywhere as well as being closed.

9.8. CURVES WITH RADIUS EQUAL TO A FUNCTION OF HARMONICS

9.8.1. Radius Equal to Sum of Harmonics
In this section, we will examine curves of the form

$$r = a + b\sin(\theta + p\pi) + c\sin(m\theta/n + q\pi)$$

The fundamental period of the curves is then $2n\pi$, and only one axis of symmetry will exist. For simplicity, only values of p and q equal to 0, 1/2, 1, or 3/2 will be

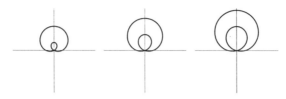

FIGURE 9.8.1. Limacon of Pascal for $a = 0.2$, $b = 0.4$, 0.6, 0.8.

FIGURE 9.8.2. Limacon of Pascal for $a = 0.5$, $b = 0.1$, 0.3, 0.5.

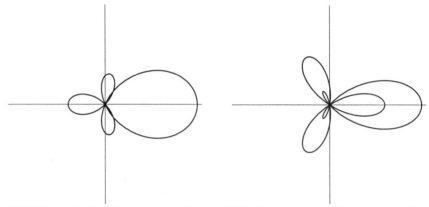

FIGURE 9.8.3.   Radius = Constant + Sum of Har-monics.

FIGURE 9.8.4.   Radius = Constant + Sum of Har-monics.

used. Clearly, more harmonics could be added; but a large variety of curves can be generated with only two harmonics.

We begin with the case where $c = 0$; this produces a curve called the *limacon of Pascal*. Letting $p = 0$ gives the curves of Figure 9.8.1 and 9.8.2 for various values of a and b. The limacon is a simple closed curve for $a > b$, forms a cusp at $(0, 0)$ whenever $a = b$, and then has a loop for $a < b$. The limiting form is a circle of radius a whenever $b = 0$. By doing the trigonometry, we find that a circle will also result when $a = 0$; in this case, its radius is $b/2$ and its center is at $b/2$ on the y axis. This circle is actually traced twice as θ goes from 0 to 2π.

Now let c be other than zero to get the sum of harmonics. First, we look at the case where $n = 1$ so that the factor on θ is simply m. Letting $p = q = 1/2$, $a = 0.2$, $b = 0.3$, $c = 0.5$, and $m = 2$ gives the curve of Figure 9.8.3. Keeping the same parameters but changing to $m = 3$ gives the curve of Figure 9.8.4.

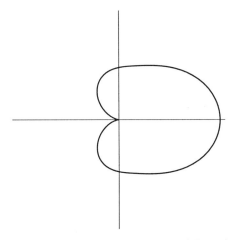

FIGURE 9.8.5. Radius = Constant + Sum of Harmonics.

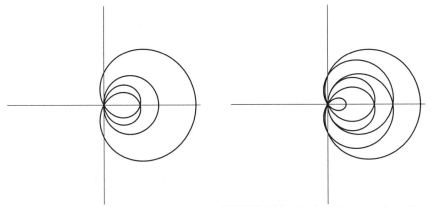

FIGURE 9.8.6. Radius = Constant + Sum of Harmonics.

FIGURE 9.8.7. Radius = Constant + Sum of Harmonics.

The number of zero crossings in the graph of r versus θ governs the number of lobes. This graph can have a maximum of m such crossings but may have less by a multiple of 2. It is possible to produce curves with no symmetry using this sum-of-harmonics form when either p or q is not 1/2, but they illustrated here. It is also important to note that certain combinations of parameters, generalized by $a \geq b + c$, will produce curves without loops. An example in Figure 9.8.5 with $p = q = 1/2$, $a = 0.5$, $b = 0.4$, $c = 0.1$, and $m = 3$ illustrates this condition in its limiting case when $a = b + c$ and a cusp is formed.

If n is greater than one, the factor m/n is a rational number. In order to plot complete curves, the range of θ will need to be $[0, 2n\pi]$. The example in Figure 9.8.6 uses $p = q = 1/2$, $a = 0.2$, $b = 0.6$, $c = 0.2$, $m/n = 1/2$. The last example in Figure 9.8.7 uses the same parameters as the curve in Figure 9.8.6 except m/n is changed to 2/3.

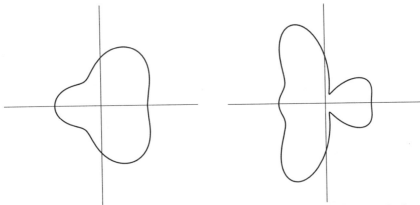

FIGURE 9.8.8.   Radius = Constant + Product of Harmonics.

FIGURE 9.8.9.   Radius = Constant + Product of Harmonics.

9.8.2.   Radius Equal to a Product of Harmonics

This section will treat curves of the form

$$r = a + b \sin^m(k\theta) \cos^n(l\theta)$$

where the range of θ is $[0, 2\pi]$. It is possible to produce curves with no symmetry from this form by using a rational number for k or l, but these will not be illustrated here. We will treat only the cases where all the parameters k, l, m, n are whole integers. The first example in Figure 9.8.8 uses $a = b = 0.5$, $k = l = n = 1$, and $m = 2$. If we increase the constant in the argument of the cosine term, it must be an odd integer to retain just one axis of symmetry. The example in Figure 9.8.9 uses the same parameters as the curve in Figure 9.8.8 except that $l = 3$ now.

On the other hand, if we change the power on the sine term to an odd integer, the constant in the argument of the cosine factor must be even to retain just one axis of symmetry. The example in Figure 9.8.10 uses $a = 0.3$, $b = 0.7$, $m = n = 1$, $k = 1$, and $l = 2$. The last example in Figure 9.8.11 uses the same parameters as the curve in Figure 9.8.10 except that $m = 3$ now.

The polar form introduced here will generate more and more complex curves as any one or more of the parameters is increased in value. Cusps and loops can be avoided by keeping $a \geq b$.

9.9.   CURVES WITH RADIUS EQUAL TO A FUNCTION OF EXPONENTIALS

9.9.1.   Simple Exponentials

The exponential function e^x was introduced in Chapter 4. The main properties of this function are the asymptotic limits at minus and plus infinity and the fact that all derivatives of an exponential function are equal to an exponential function within a constant. In this section, we will make use of these properties to produce closed curves. The basic exponential e^x does not satisfy the needs for a closed curve though; no range of x can be used such that the value of e^x is identical at both endpoints. To

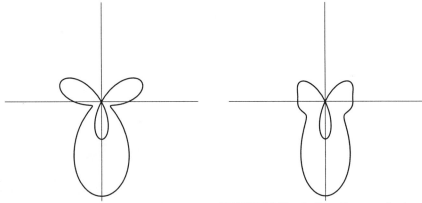

FIGURE 9.8.10. Radius = Constant + Product of Harmonics.

FIGURE 9.8.11. Radius = Constant + Product of Harmonics.

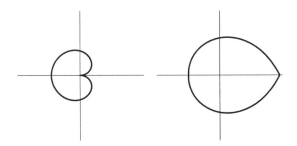

FIGURE 9.9.1. Closed curves using exponential functions.

overcome this problem and to make the function symmetric around the circle, it is necessary to work with $\exp(|\theta|)$ or $\exp(\theta^2)$ and to let the range of θ be $[-\pi, \pi]$. The result will still not be ideal because the derivative of these functions at $-\pi$ and π are equal but opposite in sign, thus giving a discontinuity in derivative even though the curve itself is continuous. This undesirable feature is greatly suppressed when the coefficient of the argument of the exponential is large enough that $dr/d\theta$ is nearly zero at the endpoints. However, in the case of $\exp(|\theta|)$, it will still be prominent at the midpoint $\theta = 0$ because this function has a discontinuous derivative there.

We start with the simple expression

$$r = a + b \exp(c|\theta|)$$

The examples in Figures 9.9.1 use $a = 0.5$, $c = -1$, and $b = \pm 0.5$. By increasing the value of b relative to a, loops can be formed. Figure 9.9.2 illustrates this by using $a = 0.3$, $c = -1$ and $b = \pm 0.7$.

If we change to the expression

$$r = a + b \exp(\theta^2)$$

hen we can achieve smoothness at $\theta = 0$. The plots in Figure 9.9.3 use the same

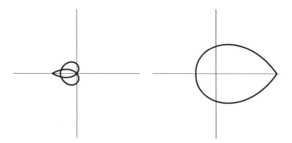

FIGURE 9.9.2. Closed curves using exponential functions.

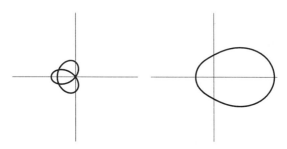

FIGURE 9.9.3. Closed curves using exponential functions.

parameters as the plots in Figure 9.9.2 except now with the θ^2 argument in the exponential function.

9.9.2. Exponentials Weighted by Powers of θ

By weighting the exponential with powers of θ, we can create some variations on the previous shapes. If the power is an odd integer, the absolute value of θ is used to retain symmetry. Recall that we use the range $-\pi < \theta < \pi$. We start with the simple expression

$$r = a + b|\theta| \exp(c|\theta|)$$

Figure 9.9.4 shows an example using $a = 0.6$, $c = -1$, and $b = \pm0.4$.

As before, we can eliminate the discontinuity in derivative at $\theta = 0$ by changing to θ^2

$$r = a + b\theta^2 \exp(c\theta^2)$$

Using the same parameters as the curves of Figure 9.9.4, the graphs now appear as they do in Figure 9.9.5. They are much smoother.

9.10. SUMMARY

This chapter has concentrated on closed curves that are non-periodic in the angular direction. Such curves were generated by several methods, including harmonic weighting of ellipses, using combinations of harmonics, and applying exponential

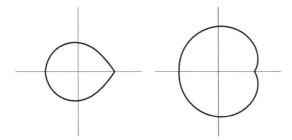

FIGURE 9.9.4. Closed curves using exponential functions.

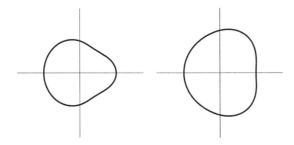

FIGURE 9.9.5. Closed curves using exponential functions.

functions. We examined a special case of trochoids which gives a non-periodic closed curve. The large and interesting family of Lissajous curves was introduced, and many variations on this type of curve were illustrated. By using the techniques of this chapter, we can produce a closed curve of nearly any desirable form to meet a particular design goal. Everything suggested here is based solely on the fundamental functions: monomials, harmonics, and exponentials.

Chapter 10

TRANSFORMATION OF CURVES

10.1. DEFINITION AND USAGE OF TRANSFORMS

10.1.1. Matrix Multiplication

Although the transformation of curves in the two-dimensional plane can be handled elegantly with complex variables, this approach requires introduction of a whole new branch of mathematics to this text and consequently will not be used. Transformation can also be treated with matrix multiplication; this, fortunately, only requires a modest introduction of concepts and is a much simpler approach without any major disadvantages for the purposes of this text. Moreover, the matrix approach is immediately programmable in available computer languages and requires no complex variable facilities. This section presents a few concepts that will aid you in understanding and using matrices.

A *scalar* is a single element. A *vector* is a linear array of elements: either numeric or symbolic. The following are three examples of vectors. Note that they can be arranged vertically or horizontally.

$$[4\,3\,7\,5] \quad \begin{bmatrix} xy \\ y^3 \\ 15y \end{bmatrix} \quad \begin{bmatrix} \cos(z) \\ \sin(z) \end{bmatrix}$$

For two vectors of equal length k, we define the *dot product* as the sum of the products of the k individual elements. For instance, the dot product of (2, 3) and (4, 1) is $2 * 4 + 3 * 1 = 11$. Note that the dot product is always a scalar. A *matrix* is a rectangular array of elements: either numeric or symbolic. It can also be described as an arrangement of vectors all of equal length. The following are examples of matrices:

$$\begin{bmatrix} 2 & 4 \\ 5 & 6 \\ 3 & 1 \end{bmatrix} \quad \begin{bmatrix} x & 2y \\ y & 3x \end{bmatrix} \quad \begin{bmatrix} \cos(a) & -\sin(a) \\ \sin(a) & \cos(a) \end{bmatrix} \quad \begin{bmatrix} 1 & 5 & 9 \\ 2 & 1 & 8 \end{bmatrix}$$

We generalize the shape of a matrix by saying it has n rows by m columns or, briefly, is "n by m." Note that vectors can be considered as matrices of 1 by m (row vector) or n by 1 (column vector). An element within a matrix C is referred to as c_{nm} where n and m are the row and column numbers. We define matrix multiplication as follows. Given two matrices A and B, the matrices can be multiplied together to get $C = AB$ if and only if the number of columns of A is equal to the number of rows of B. Provided this is true, the element c_{nm} is formed by the dot product of the nth row of A and the mth column of B. An example should help to clarify this.

$$\begin{bmatrix} 2 & 4 \\ 3 & 2 \end{bmatrix} \begin{bmatrix} 1 & 3 \\ 2 & 5 \end{bmatrix} = \begin{bmatrix} 10 & 26 \\ 7 & 19 \end{bmatrix}$$

In applying transformations to curves, the vectors to be used will all be of length 2, which is the number of Cartesian or polar coordinates for a point in the plane.

(Note that such vectors can be considered matrices of 1 by 2 or 2 by 1; they are often called "coordinate vectors.") The corresponding matrices must then be 2 by 2 to produce another transformed pair of coordinates. Thus, the calculations required for transforming curves in this way are trivial. For each coordinate, two multiplications and a single sum are all that is required.

10.1.2. General Transformation Matrix

Using the definitions of the preceding section, we write the general transformation of Cartesian coordinates in a plane as

$$\begin{bmatrix} x' \\ y' \end{bmatrix} = \begin{bmatrix} a & b \\ c & d \end{bmatrix} \begin{bmatrix} x \\ y \end{bmatrix}$$

A similar relation can be written for other coordinates systems such as the polar. Let the matrix be written as $\mathbf{T}$. If the coordinate pair (x, y) is denoted by p and the coordinate pair (x', y') by p', then $p' = \mathbf{T}p$. The values of a, b, c, and d are, for the moment, undefined. The above matrix notation is equivalent to the equations

$$x' = ax + by$$
$$y' = cx + dy$$

We should understand that these equations are implied when we speak of transforming coordinates by a matrix $\mathbf{T}$. To produce an individual p', the implemented method must compute the two equations above even though it is denoted simply as $p' = \mathbf{T}p$. To produce all the points p' associated with the points p, these equations are invoked repeatedly.

By writing the equations, we see that a transformation is a function, or pair of functions in this case, that maps one set of points to another set. The transformation is called *one-to-one* if, for every point p, there exists a unique point p'. The transformation is called *onto* if, for every point p, the p' is identical to p. It is important to recognize the *fixed points* of a transform as being those points where $p' = p$. In the case of $\mathbf{T}$ being onto, all points are fixed; but in general there will be only one or a limited set of such points. Clearly, by the definition above, $(0, 0)$ is always a fixed point of every transformation.

Some classes of transformations are so common and intuitive that they have acquired special names. We first mention the *linear transform*, for which all the elements of $\mathbf{T}$ are real numbers. The linear transform can stretch (or compress) a two-dimensional figure in either direction, flip it about either axis, and also rotate it. One, two, or all three of these operations may be contained in a particular linear transform. Nevertheless, the result is always easily associated with the original visually. A *simple* transform maps a simple closed curve into another simple closed curve. Another class of transform is called *conformal*; these tranforms have the property that angles between intersecting lines are not changed by transforming the lines.

10.1.3. Transformation Extended to Include Translation

In working with curves in the two-dimensional plane, it is often desirable to simply shift them to a new position without changing them in any other way. This is called *translation* and is represented by the following equations

$$x' = x + e$$
$$y' = y + f$$

where e and f are distances along the x and y axes, respectively. If the vector of translation distances (e, f) is denoted by t, then $p' = p + t$ describes translation. Sometimes it is necessary to apply both the matrix transformation given in the section above and a translation to achieve the proper appearance and position of a curve. Can these two operations be written together succinctly? If the matrix **T** is extended by one column with the column vector t and p is extended by an element equal to unity, then the following transformation achieves this:

$$\begin{bmatrix} x' \\ y' \end{bmatrix} = \begin{bmatrix} a & b & e \\ c & d & f \end{bmatrix} \begin{bmatrix} x \\ y \\ 1 \end{bmatrix}$$

Recall that, for the 2-by-2 matrix **T** introduced in the previous section, any **T** for which all the elements are real numbers is called a linear transform. For the extended **T** introduced here, when all the elements are real numbers, it is called an *affine transformation*. The above matrix notation is equivalent to the equations

$$x' = ax + by + e$$
$$y' = cx + dy + f$$

The advantage of the combined transformation is that it is compactly written and easily programmable. We can handle both translation alone and an ordinary linear transformation without translation as special cases of this combined transformation.

For the examples of this text, we will ignore translation. The purpose here is merely to show that it can be easily incorporated when the need arises. Thus, all of the transformation matrices presented here will be 2-by-2.

10.1.4. Compound Transformations

One of the nice properties of transformations as defined here is that they can be concatenated to make a compound transform. If **S** and **T** are transforms, then **R** = **ST**, formed by matrix multiplication, is a valid transform. Thus, if we know the form of the particular transforms which individually produce certain effects, then their multiplication will produce the concatenation of these operations. We must realize, however, that matrix multiplication does not have a commutative property. Thus, **ST** is not identical to **TS** in most cases. This point can be illustrated by a simple example. Take a line at 45 degrees between the x and y axes as shown in Figure 10.1.1 on the left. The middle part of this figure shows this line after reflection about the y axis and a clockwise rotation of 30 degrees. The right-hand part of the figure shows this line after reversing the order of applying these two transformations.

In practice, the order of the matrices composing the compound transformation follows from the logical definition of the sequence of desired operations. It is only necessary to ensure that this order is preserved when transcribing the matrices to the programming language being used.

FIGURE 10.1.1.   Effect of compounding transformations(see text).

10.1.5.   Effect of Transformations on Symmetry

Recall that use of the general transformation matrix is equivalent to the following two equations:

$$x' = ax + by$$
$$y' = cx + dy$$

It would be useful to know how this transformation affects the symmetry of curves. What, specifically, are the requirements on the factors a, b, c, and d to preserve symmetry or antisymmetry? Let us first discuss the case of a function that is symmetric about the y axis: $f(x) = f(-x)$. (To be correct, we must consider each of the multivalued branches of a function individually.) Looking at $y' = cx + dy$, it is the sum of two terms. If the transformed function is to be symmetric about the y axis, then both terms in $cx + dy$ must be symmetric. Because we already assumed y to be symmetric, it can only remain symmetric if d is composed of one or more of these factors: 1) a constant (including zero), 2) any power of y, and 3) an even power of x. On the other hand, x on either side of the y axis is antisymmetric. Therefore, to make cx symmetric, c must either be zero or be composed of an odd power of x and one or both of these factors: 1) a constant or 2) any power of y. Furthermore, to ensure that y' is symmetric, x' must be antisymmetric. Similar reasoning leads to the requirement that b contain an odd power of x and one or both of these factors: 1) a constant or 2) any power of y. It also leads to the requirement that a is composed of one or more of these factors: 1) a constant (including zero), 2) an even power of x, or 3) any power of y.

The description of the symmetry requirements is easier to grasp when illustrated with some examples. Consider the curve $\exp(-x^2)$ which is symmetric about the y axis (see Figure 10.1.2). Parametrically, this curve can be expressed as

$$x = t$$
$$y = \exp(-t^2)$$

In Figures 10.1.3 to 10.1.7, we apply various transformations to this curve each given by the matrix form. We should be able to quickly verify that the above analysis predicts the effect of the transformation on the symmetry. In the captions of these figures, the notation $\mathbf{T} = [(a, b), (c, d)]$ is equivalent to

$$\mathbf{T} = \begin{bmatrix} a & b \\ c & d \end{bmatrix}$$

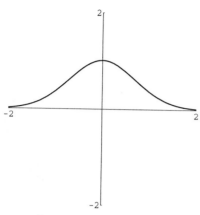

FIGURE 10.1.2. Graph of exp($-x^2$).

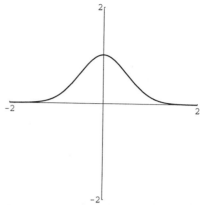

FIGURE 10.1.3. Transformation of exp($-x^2$) by **T** = [(1, 0), (0, y)].

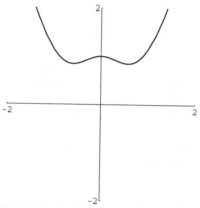

FIGURE 10.1.4. Transformation of exp($-x^2$) by **T** = [(1, 0), (x, y)].

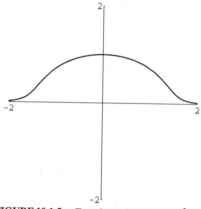

FIGURE 10.1.5. Transformation of exp($-x^2$) by **T** = [(1, x), (0, 1)].

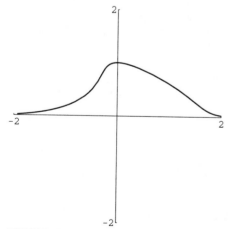

FIGURE 10.1.6. Transformation of exp($-x^2$) by **T** = [(1, x^2), (0, 1)].

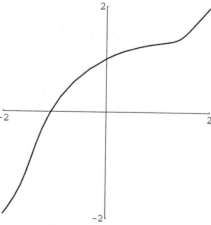

FIGURE 10.1.7. Transformation of exp($-x^2$) by **T** = [(1, x), (1, 1)].

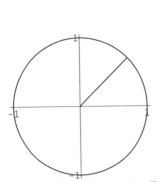

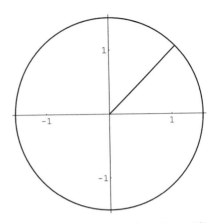

FIGURE 10.2.1.   Curves for showing effects of linear transformations.

FIGURE 10.2.2.   Simple scaling with $a = 3/2$.

If the original function is antisymmetric about the y axis and we wish to preserve this property after transformation, then the analysis leads to similar requirements. Only "even" is replaced with "odd" and vice-versa in the above analysis for symmetric curves. It should be clear that symmetry and antisymmetry about the x axis can be handled with the same analysis when the roles of x and y are reversed. We can use the results here to design a transformation such that a certain symmetry or antisymmetry is preserved at the same time that other desired effects on the curve are achieved.

10.2.   LINEAR TRANSFORMATIONS ON CLOSED CURVES IN CARTESIAN COORDINATES

10.2.1.   Introduction
This section describes and illustrates linear transformations of curves plotted in Cartesian coordinates. Linear transformations can achieve many of the desired effects on curves, especially when they are concatonated. We adopt a standard set of curves that will be used in the illustration of all such transformations. This set comprises a circle and a diagonal line from the origin, bisecting the axes, and ending on the circle as shown in Figure 10.2.1. In the following subsections, we treat all the special cases of linear transformations normally of interest and illustrate their effect on this standard set of curves. One feature to notice with all the linear transformations to follow is that the straight line segment always remains straight after transformation.

10.2.2.   Simple Scaling
This transformation is given by

$$\mathbf{T} = \begin{bmatrix} a & 0 \\ 0 & a \end{bmatrix}$$

It gives a uniform, isotropic magnification, or compression, to a curve. The example in Figure 10.2.2 is for $a = 3/2$.

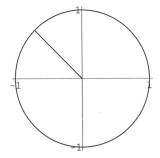

FIGURE 10.2.3. Reflection about *y* axis.

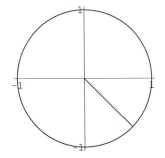

FIGURE 10.2.4. Reflection about *x* axis.

10.2.3. Reflection about the *Y* Axis
This transformation is given simply by

$$T = \begin{bmatrix} -1 & 0 \\ 0 & 1 \end{bmatrix}$$

The effect of this is shown in Figure 10.2.3.

10.2.4. Reflection about the *X* Axis
This transformation is given simply by

$$T = \begin{bmatrix} 1 & 0 \\ 0 & -1 \end{bmatrix}$$

The effect of this is shown in Figure 10.2.4.

10.2.5. Reflection through the Origin
This transformation is given simply by

$$T = \begin{bmatrix} -1 & 0 \\ 0 & -1 \end{bmatrix}$$

Figure 10.2.5 shows the effect of this transformation.

10.2.6. Rotation by an Angle α
This transformation is given by

$$T = \begin{bmatrix} \cos \alpha & -\sin \alpha \\ \sin \alpha & \cos \alpha \end{bmatrix}$$

The angle α is positive in the counterclockwise sense and is measured relative to the positive *x* axis. We can effect a clockwise rotation by using a negative angle. The example in Figure 10.2.6 uses $\alpha = 30$ degrees ($\pi/6$).

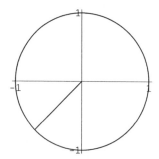

FIGURE 10.2.5. Reflection through the origin.

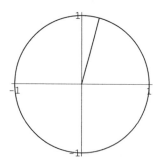

FIGURE 10.2.6. Rotation by 30 degrees.

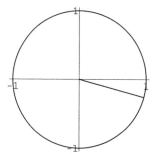

FIGURE 10.2.7. Reflection about a line at 15 degrees.

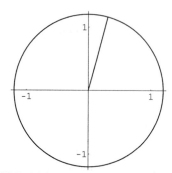

FIGURE 10.2.8. Similitude with $\alpha = 30$ degrees and scaling = 6/5.

10.2.7.   Reflection about a Line at an Angle α

This transformation is given by

$$\mathbf{T} = \begin{bmatrix} \cos 2\alpha & -\sin 2\alpha \\ \sin 2\alpha & \cos 2\alpha \end{bmatrix}$$

The angle α is positive in the counterclockwise sense and is measured relative to the positive x axis. The example in Figure 10.2.7 uses $\alpha = 15$ degrees ($\pi/12$).

10.2.8.   Similitude

A *similitude* is a restrictive linear transformation in that it must stretch (or compress) the figure equally in both directions to remain undistorted while rotating clockwise through the angle α. It is given by

$$\mathbf{T} = \begin{bmatrix} a\cos\alpha & -a\sin\alpha \\ a\sin\alpha & a\cos\alpha \end{bmatrix}$$

Note that this is actually a compound transformation because it is the concatenation of simple scaling and rotation. The example in Figure 10.2.8 uses $\alpha = 30$ degrees ($\pi/6$) and $a = 6/5$.

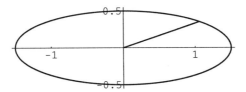

FIGURE 10.2.9. Arbitrary scaling with $a = 3/2$ and $d = 1/2$.

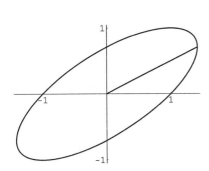

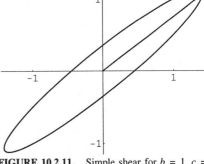

FIGURE 10.2.10. Simple shear for $b = 1$, $c = 0$.

FIGURE 10.2.11. Simple shear for $b = 1$, $c = 1/2$.

10.2.9. Arbitrary Scaling
This transformation is given by

$$\mathbf{T} = \begin{bmatrix} a & 0 \\ 0 & d \end{bmatrix}$$

This stretches or compresses the figure along the x and y axes according to the values a and d, respectively. The example in Figure 10.2.9 is for $a = 3/2$ and $d = 1/2$.

10.2.10. Simple Shear
This transformation is given by

$$\mathbf{T} = \begin{bmatrix} 1 & b \\ c & 1 \end{bmatrix}$$

This transformation shears the figure parallel to the x axis for $b > 0$ and parallel to the y axis for $c > 0$. The shear displacement increases linearly in the direction perpendicular to the axis starting from zero at the axis itself. The first example in Figure 10.2.10 shows the effect for $b = 1$ and $c = 0$. Note that the intercepts on the x axis remain identical to those of the original circle. Another example in Figure 10.2.11 uses $b = 1$ and $c = 1/2$. Now the x intercepts are also changed.

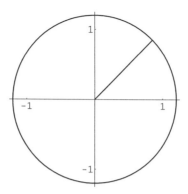

FIGURE 10.3.1. Simple scaling with $a = 6/5$.

10.3. LINEAR TRANSFORMATIONS ON CLOSED CURVES IN POLAR COORDINATES

10.3.1. Introduction

If the coordinates are taken to be the polar coordinates (r, θ), then we can define the following 2-by-2 transformation matrix:

$$\begin{bmatrix} r' \\ \theta' \end{bmatrix} = \begin{bmatrix} a & b \\ c & d \end{bmatrix} \begin{bmatrix} r \\ \theta \end{bmatrix}$$

Although the actual numbers in the transform matrix may be identical for both Cartesian and polar coordinates, the effect may be greatly different. Some of the same types of transformations presented for Cartesian coordinates are possible with polar coordinates also. The preservation of the linearity of the straight line is no longer guaranteed by this transformation though. A few of the useful variations of such a transformation are given in the following subsections.

10.3.2. Simple Scaling

This transformation is given by

$$\mathbf{T} = \begin{bmatrix} a & 0 \\ 0 & 1 \end{bmatrix}$$

This will uniformly and isotropically magnify or compress the figure. The example in Figure 10.3.1 shows the result for $a = 6/5$.

10.3.3. Reflection through the Origin

This transformation is given by

$$\mathbf{T} = \begin{bmatrix} -1 & 0 \\ 0 & 1 \end{bmatrix}$$

Figure 10.3.2 shows the effect.

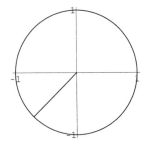

FIGURE 10.3.2. Reflection through the origin.

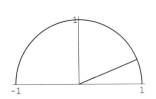

FIGURE 10.3.3. Angular scaling with $d = 1/2$.

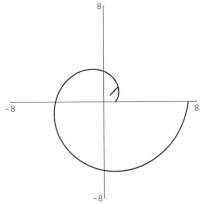

FIGURE 10.3.4. Arbitrary scaling with $a = 1$, $b = 1, c = 0, d = 1$.

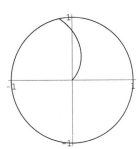

FIGURE 10.3.5. Arbitrary scaling with $a = 1$, $b = 0, c = 1, d = 1$.

10.3.4. Angular Scaling
This transformation is given by

$$\mathbf{T} = \begin{bmatrix} 1 & 0 \\ 0 & d \end{bmatrix}$$

All angles are decreased by the factor d. Thus, the total angular range will be reduced or enlarged by that factor; and the original figure will be contained within the new angular range. The example in Figure 10.3.3 uses $d = 1/2$.

10.3.5. Arbitrary Scaling
This transformation is given by

$$\mathbf{T} = \begin{bmatrix} a & b \\ c & d \end{bmatrix}$$

The example in Figure 10.3.4 uses $a = b = d = 1$ and $c = 0$. Because r is increasing with θ, a spiral effect is produced. Another example in Figure 10.3.5 uses $a = c = d = 1$ and $b = 0$. Note that the straight line is now curved.

FIGURE 10.4.1. Curves used to show effects of nonlinear transformations.

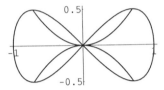

FIGURE 10.4.2. Twisting transformation about X axis.

10.4. NONLINEAR TRANSFORMATIONS ON CLOSED CURVES IN CARTESIAN COORDINATES

10.4.1. Introduction

For linear transformations, the variables representing the coordinates never appeared in the transformation matrix itself. A *nonlinear transformation* will have one or both variables in one or more of the four elements of the transformation matrix. The possible cases are unlimited; therefore, we can only look at a few of the simpler ones here. To help clarify the effect of the nonlinear transformations, we will use a somewhat more complex form as the basis for illustrating the transformations. This will consist of the circle again but with two diagonal lines at 45 degrees to the axes as shown in Figure 10.4.1.

10.4.2. Twisting about X Axis

This transformation is given by

$$\mathbf{T} = \begin{bmatrix} 1 & 0 \\ 0 & dx \end{bmatrix}$$

Because $y' = dx$, the circle will be forced to pass through the origin, making a twist about the x axis. The diagonal lines ($\sim x$) are deformed into parabolas ($\sim x^2$). The example in Figure 10.4.2 shows the result for $d = 1$.

10.4.3. Twisting about Y Axis

This transformation is given by

$$\mathbf{T} = \begin{bmatrix} ay & 0 \\ 0 & 1 \end{bmatrix}$$

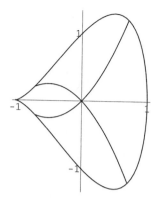

FIGURE 10.4.3. Compression and expansion on opposite side of Y axis.

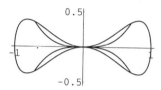

FIGURE 10.4.4. Expansion proportional to x^2.

This is basically the same transformation as the previous one except that the twist is about the y axis. The example in Figure 10.4.2 would be rotated by 90 degrees if we used $a = 1$ here.

10.4.4. Compression and Expansion on Opposite Sides of Y Axis
This transformation is given by

$$\mathbf{T} = \begin{bmatrix} 1 & 0 \\ cy & 1 \end{bmatrix}$$

The example in Figure 10.4.3 shows the result for $c = 1$; the result for $c = -1$ is the reflection of this example about the y axis.

10.4.5. Compression and Expansion on Opposite Sides of X Axis
This transformation is given by

$$\mathbf{T} = \begin{bmatrix} 1 & bx \\ 0 & 1 \end{bmatrix}$$

The result will be identical to that for the transformation in Section 10.4.4 except rotated by 90 degrees.

10.4.6. Y Expansion Proportional to X^2
This transformation is given by

$$\mathbf{T} = \begin{bmatrix} 1 & 0 \\ 0 & dx^2 \end{bmatrix}$$

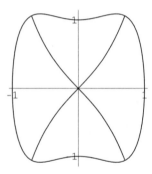

FIGURE 10.4.5. Expansion proportional to $1 + cx^2$.

This is similar to the transformation in Section 10.4.2; now the circle does not twist but merely touches the origin due to the x^2 factor, as shown in Figure 10.4.4.

10.4.7. *X* Expansion Proportional to Y^2

This transformation is given by

$$\mathbf{T} = \begin{bmatrix} ay^2 & 0 \\ 0 & 1 \end{bmatrix}$$

The result will be the same as that for the transformation in Section 10.4.6 except for a rotation of 90 degrees.

10.4.8. *Y* Expansion Proportional to $1 + CX^2$

This transformation is given by

$$\mathbf{T} = \begin{bmatrix} 1 & 0 \\ cxy & 1 \end{bmatrix}$$

The example in Figure 10.4.5 is for $c = 1$.

10.4.9. *X* Expansion Proportional to $1 + BY^2$

This transformation is given by

$$\mathbf{T} = \begin{bmatrix} 1 & bxy \\ 0 & 1 \end{bmatrix}$$

The result will be identical to that for the transformation in Section 10.4.8 except for a rotation of 90 degrees.

10.4.10. Stretching along Lines Which Bisect the Axes

This transformation is given by

$$\mathbf{T} = \begin{bmatrix} 1 & bxy \\ cxy & 1 \end{bmatrix}$$

The example in Figure 10.4.6 is for $b = c = 1$.

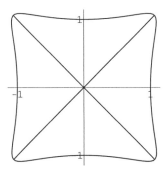

FIGURE 10.4.6. Stretching along lines which bisect the axes.

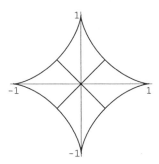

FIGURE 10.4.7. Stretching along the axes.

10.4.11. Stretching along the Axes
This transformation is given by

$$\mathbf{T} = \begin{bmatrix} a|x|^n & 0 \\ 0 & d|y|^n \end{bmatrix}$$

The exponent n is assumed to be an integer. Note that the absolute value is not required when n is an even integer. The example in Figure 10.4.7 uses $a = 1$ and $d = 1$ and $n = 2$.

10.5. LINEAR TRANSFORMATIONS ON HARMONIC FUNCTIONS

10.5.1. Introduction
Thus far we have studied the effect of transformations on simple curves comprising lines and a circle. Another fundamental curve shape is the sine wave, and it is useful to see the effect of various transformations on this curve. The sine wave can be expressed parametrically as

$$x = t$$
$$y = \sin(t + \phi)$$

We will look at only those cases where $\phi = 0$ (sine function) or $\phi = \pi$ (cosine function). This section will treat only linear transformations; recall that such transformations

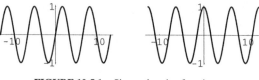

FIGURE 10.5.1.   Sine and cosine functions.

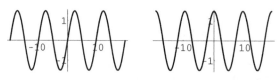

FIGURE 10.5.2.   Simple scaling of sine and cosine functions.

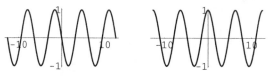

FIGURE 10.5.3.   Reflection about the *Y* axis of sine and cosine functions.

FIGURE 10.5.4.   Reflection about the *X* axis of sine and cosine functions.

have only constants in the matrix **T**. These transformations are identical to those introduced in Section 10.2 except that they now operate on harmonic functions. Figure 10.5.1 shows the two basic harmonic functions in their original form.

10.5.2.   Simple Scaling
This transformation is identical in form to that in Section 10.2.2. The example in Figure 10.5.2 is for $a = 3/2$. The result of this transformation are curves that are not visibly different from the original functions, but they are stretched along both directions. Using different constants for the x and y scaling is a natural extension.

10.5.3.   Reflection about the *Y* Axis
This transformation is identical in form to that in Section 10.2.3. The example is shown in Figure 10.5.3. The sine function is affected by this transformation; but the cosine function, due to its symmetry, is not affected.

10.5.4.   Reflection about the *X* Axis
This transformation is identical in form to that in Section 10.2.4. The example is shown in Figure 10.5.4. For this transformation, both the sine and cosine functions are visibly affected.

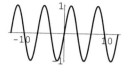 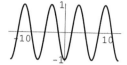

FIGURE 10.5.5. Reflection through the origin of sine and cosine functions.

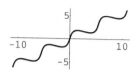 

FIGURE 10.5.6. Rotation by 30 degrees of sine and cosine functions.

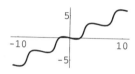 

FIGURE 10.5.7. Reflection about 15-degree line of sine and cosine functions.

10.5.5. Reflection through the Origin

This transformation is identical in form to that in Section 10.2.5. The example is shown in Figure 10.5.5. In this case, the sine wave is unaffected due to its antisymmetry, and the cosine wave is turned upside down.

10.5.6. Rotation

This transformation is identical in form to that in Section 10.2.6. The angle α is positive in the counterclockwise sense and is measured relative to the positive x axis. We can effect a clockwise rotation by using a negative angle. The example in Figure 10.5.6 is for $\alpha = 30$ degrees ($\pi/6$). Note that the antisymmetry of the sine function is preserved, and the cosine function lacks any symmetry about the x or y axes after this transformation.

10.5.7. Reflection about a Line at an Angle α

This transformation is identical in form to that in Section 10.2.7. The angle α is positive in the counterclockwise sense and is measured relative to the positive x axis. The example in Figure 10.5.7 uses $\alpha = 15$ degrees ($\pi/12$). Again, as with the rotation transformation, the sine function retains its antisymmetry, and the cosine function loses its symmetry.

10.5.8. Similitude

This transformation is identical in form to that in Section 10.2.8. The example in Figure 10.5.8 uses $\alpha = 30$ degrees ($\pi/6$) and $a = 6/5$.

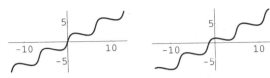

FIGURE 10.5.8. Similitude (α = 30 degrees and a = 6/5) of sine and cosine functions.

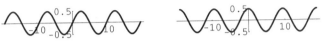

FIGURE 10.5.9. Arbitrary scaling (a = 3/2, d = 1/2) of sine and cosine functions.

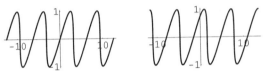

FIGURE 10.5.10. Simple shear (b = 1, c = 0) of sine and cosine functions.

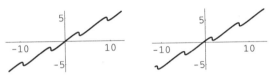

FIGURE 10.5.11. Simple shear (b = 1, c = 1/2) of sine and cosine functions.

10.5.9.   Arbitrary Scaling

This transformation is identical in form to that in Section 10.2.9. The example in Figure 10.5.9 is for a = 3/2 and d = 1/2.

10.5.10.   Simple Shear

This transformation is identical in form to that in 10.2.10. The first example in Figure 10.5.10 shows the effect for b = 1 and c = 0. The effect is to make both functions "lean" in the forward (+x) direction above the x axis and backward below the x axis. A second example in Figure 10.5.11 is for b = 1 and c = 1/2.

10.6.   NONLINEAR TRANSFORMATIONS ON HARMONIC FUNCTIONS

10.6.1.   Introduction

Recall that "nonlinear" transformations have a factor with a power of x or y in one or more of the elements of the matrix **T**. Such forms, limitless in number, can produce some interesting effects on sine or cosine waves. We will consider a few of these in this section. We will be looking specifically at

$$\mathbf{T} = \begin{bmatrix} a & b \\ c & d \end{bmatrix}$$

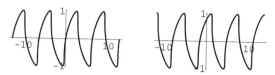

FIGURE 10.6.1. Quadratic bending of sine and cosine functions.

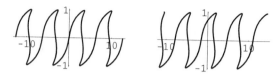

FIGURE 10.6.2. Cubic bending of sine and cosine functions.

limited to where $c = 0$ and $d = 1$, while a and b are arbitrary functions of x. This gives the transformation equations

$$x' = ax + by$$
$$y' = y$$

so that y is unchanged, while x is changed to a function of x and y. Some of the results here can be duplicated by using the methods of Chapter 5.

10.6.2. Quadratic Bending
This transformation is given by

$$\mathbf{T} = \begin{bmatrix} 1 & ky \\ 0 & 1 \end{bmatrix}$$

The example in Figure 10.6.1 is for $k = 1$. Compare this result with that for the example of simple shear in Figure 10.5.10. The bending there was linear in y, but here it is quadratic. The quadratic effect makes the sine wave "lean" forward both above and below the x axis.

10.6.3. Cubic Bending
This transformation is given by

$$\mathbf{T} = \begin{bmatrix} 1 & ky^2 \\ 0 & 1 \end{bmatrix}$$

The example in Figure 10.6.2 is for $k = 2$. Here the bending is forward above the x axis and backward below it.

10.6.4. Quartic Bending
This transformation is given by

$$\mathbf{T} = \begin{bmatrix} 1 & ky^3 \\ 0 & 1 \end{bmatrix}$$

The example in Figure 10.6.3 is for $k = 2$. This is similar to the quadratic bending in Figure 10.6.1 but further enhanced.

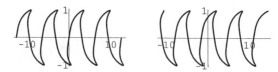

FIGURE 10.6.3. Quartic bending of sine and cosine functions.

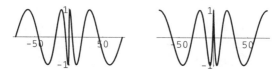

FIGURE 10.6.4. Linear X expansion of sine and cosine functions.

10.6.5. Linear Expansion of X

This transformation is given by

$$\mathbf{T} = \begin{bmatrix} k|x| & 0 \\ 0 & 1 \end{bmatrix}$$

We use the absolute value to have the effect be identical for both the negative and positive domains of x. The example in Figure 10.6.4 is for $k = 1/2$. Note that the transformed sine wave retains its antisymmetry about the y axis and that the transformed cosine wave retains its symmetry.

10.7. BENDING TRANSFORMATION

An interesting transformation not included in the above classes is a "bending" transformation. We will consider this transformation acting on the ellipse, which is given by the following parametric equations:

$$x = a\cos(t)$$
$$y = b\sin(t)$$

The particular transformation will be given by the following equations:

$$x' = x\cos(cx) + y\sin(cx)$$
$$y' = -x\sin(cx) + y\cos(cx)$$

In matrix form, this is

$$\mathbf{T} = \begin{bmatrix} \cos(cx) & \sin(cx) \\ -\sin(cx) & \cos(cx) \end{bmatrix}$$

This appears to be similar in form to the rotation matrix of Section 10.2.6, but here the argument of the sine and cosine functions is not a constant. It is designed to preserve symmetry about the y axis but not about the x axis. We apply this transformation using $a = 1$, $b = 1/2$, and $c = 1/4$ in Figure 10.7.1. Figure 10.7.2 uses $a = 1$ and

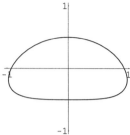

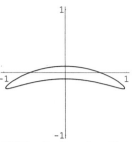

FIGURE 10.7.1. Bending of an ellipse ($a = 1$, $b = 0.5$, $c = 0.25$).

FIGURE 10.7.2. Bending of an ellipse ($a = 1$, $b = 0.1$, $c = 0.25$).

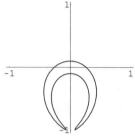

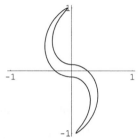

FIGURE 10.7.3. Bending of an ellipse ($a = 1$, $b = 0.1$, $c = 1.5$).

FIGURE 10.7.4. Antisymmetric bending of an ellipse.

$c = 1/4$ again, but b is now set to 1/10. The last example in Figure 10.7.3 uses $a = 1$ and $b = 1/10$ again, but c is now set to 3/2.

A different effect can be obtained by using the absolute value of x as the argument of the sine and cosine functions:

$$x' = x\cos(c|x|) + y\sin(c|x|)$$
$$y' = -x\sin(c|x|) + y\cos(c|x|)$$

In matrix form, this is

$$\mathbf{T} = \begin{bmatrix} \cos(c|x|) & \sin(c|x|) \\ -\sin(c|x|) & \cos(c|x|) \end{bmatrix}$$

This will cause the bend to go upward on one side of the y axis and downward on the other. Applying this to the example of Figure 10.7.3 with the same parameters produces the antisymmetric curve in Figure 10.7.4.

We can make similar effects in the x direction rather than the y direction by changing the argument of the sine and cosine functions from x to y in all of the above. Variations can be made by using a power of x other than unity in these arguments.

10.8. LOOPING TRANSFORMATION

Another interesting transformation not covered in any of the above sections is the "looping" transformation. It can be expressed compactly if we recall the relation

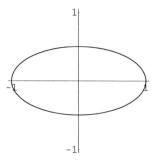

FIGURE 10.8.1. Circle transformed to an ellipse.

between Cartesian and polar coordinates:

$$x = r\cos(\theta)$$
$$y = r\sin(\theta)$$

The particular transformation will be given by the following equations:

$$x' = ax + b\sin(n\theta)$$
$$y' = dy + c\cos(n\theta)$$

In matrix form, this is

$$\mathbf{T} = \begin{bmatrix} a & b\sin(n\theta)/y \\ c\cos(n\theta)/x & d \end{bmatrix}$$

The angular variable θ still remains in this expression for $\mathbf{T}$; however, it can be eliminated by noting that both $\cos(n\theta)$ and $\sin(n\theta)$ can be expanded in a polynomial series involving powers of $\cos(\theta)$ and $\sin(\theta)$. The following relations enable us to recursively calculate this series for any factor n.

$$\sin(2\theta) = 2\sin(\theta)\cos(\theta)$$
$$\cos(2\theta) = \cos^2(\theta) - \sin^2(\theta)$$
$$\sin(n\theta) = 2\sin[(n-1)\theta]\cos(\theta) - \sin[(n-2)\theta]$$
$$\cos(n\theta) = 2\cos[(n-1)\theta]\cos(\theta) - \cos[(n-2)\theta]$$

For $n = 1$, the transformed parametric equations become

$$x' = ax + by/r$$
$$y' = cx/r + dy$$

When $b = c = 0$, this transforms the circle (r = constant) to an ellipse with semimajor axes of length a and b. The example in Figure 10.8.1 uses $a = 1$ and $d = 1/2$. When b and c are nonzero, this effects the symmetry about the x and y axes in addition to the scaling due to a and d. If the original figure is a circle, the resultant figure is

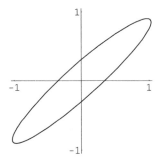

FIGURE 10.8.2. Circle transformed to pseudo-ellipse.

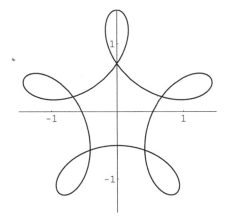

FIGURE 10.8.3. Circle transformed to hypotrochoid.

not, however, an ellipse except in a very special case when the coefficients are equal to those that result when the transformations of 10.2.9 and 10.2.6 were applied in succession. Figure 10.8.2 uses this transform again on a circle with $a = 1$, $b = 1/4$, $c = 3/4$, and $d = 1/2$.

We suggested the current transformation as a "looping" transformation but have not yet produced them with $n = 1$. To deform the circle into a figure with loops, we must use $n > 1$ to get $n + 1$ loops exactly. When $a = d$ and $b = c$, the form is actually the hypotrochoid of Section 6.4.1. The loops degenerate to cusps when b and c are each equal to a or d divided by n. The hypotrochoid form is illustrated in Figure 10.8.3 with $a = d = 1$, $b = c = 0.5$, and $n = 4$. The more general form is illustrated in Figure 10.8.4 with $a = 1$, $d = 0.5$, $b = c = 0.3$, and $n = 4$ again.

Finally, the most general case uses different values for all four coefficients. This is illustrated in Figure 10.8.5, again for $n = 4$, with $a = 1.0$, $b = 0.3$, $c = 0.7$, and $d = 0.5$.

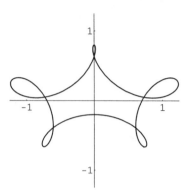

FIGURE 10.8.4. Circle transformed to pseudo-hypotrochoid.

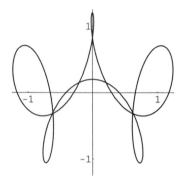

FIGURE 10.8.5. Circle transformed to pseudo-hypotrochoid.

10.9.   SUMMARY

This chapter has shown how we can transform a curve into another shape by apply-ing matrix transformations. These transformations are particularly valuable "tools" that you should know and apply. The transformations can be classed as linear or nonlinear, depending on whether the matrix contains all constants or not. The linear transformations are simple in form, and it is fairly easy to visualize their effect just from the matrix form itself. Those linear transformations introduced in this chapter are repeated here in tabular format.

Basic Linear Transformations

simple scaling $\qquad\qquad\qquad \mathbf{T} = \begin{bmatrix} a & 0 \\ 0 & a \end{bmatrix}$

reflection about Y axis $\qquad\quad \mathbf{T} = \begin{bmatrix} -1 & 0 \\ 0 & 1 \end{bmatrix}$

reflection about X axis $\qquad\quad \mathbf{T} = \begin{bmatrix} 1 & 0 \\ 0 & -1 \end{bmatrix}$

reflection through origin
$$T = \begin{bmatrix} -1 & 0 \\ 0 & -1 \end{bmatrix}$$

rotation by angle α
$$T = \begin{bmatrix} \cos(\alpha) & -\sin(\alpha) \\ \sin(\alpha) & \cos(\alpha) \end{bmatrix}$$

reflection about line at angle α
$$T = \begin{bmatrix} \cos(2\alpha) & -\sin(2\alpha) \\ \sin(2\alpha) & \cos(2\alpha) \end{bmatrix}$$

similitude
$$T = \begin{bmatrix} a\cos(\alpha) & -a\sin(\alpha) \\ a\sin(\alpha) & a\cos(\alpha) \end{bmatrix}$$

arbitrary scaling
$$T = \begin{bmatrix} a & 0 \\ 0 & d \end{bmatrix}$$

simple shear
$$T = \begin{bmatrix} 1 & b \\ c & 1 \end{bmatrix}$$

Once we learn the nature and form of a particular transformation we can apply it with appropriate constants or variables to produce a desired effect. One of the powerful features of matrix transformations is that they can be concatonated together to achieve the desired effect in controlled steps. Although the examples here were limited to lines, circles, and harmonic functions, the transformations are applicable to all curves.

Nonlinear transformations can produce more extreme and complex deformations of curves than linear ones. As examples of useful transformations of nonlinear type, matrix forms were introduced to bend and twist curves, to cause them to have loops, and to produce nonlinear scaling along one or other of the axes.

Chapter 11

COMPLEX CURVES FROM POLYNOMIALS

11.1. BASIS FUNCTIONS

In this chapter and the ones to follow, we will attempt to approximate given curves $y(x)$, which may be complex, with a set of simpler functions according to the scheme:

$$y(x) \approx a_0 + a_1 f_1(x) + a_2 f_2(x) + \ldots + a_n f_n(x)$$

The $y(x)$ may be expressed functionally or may be given as a finite set of points at discrete values of x. We assume that the functions $f_i(x)$ are all similar; for instance, all polynomials or all harmonics. In the case of polynomials, the $f_i(x)$ will be monomials of increasing power. In selecting a set of functions to approximate $y(x)$, it is said that they are a *basis*; thus the f_i are the basis functions. It is desirable that the basis be *compact*. This means that the number of functions used to approximate $y(x)$ should be as small as possible within the criterion of fitting. Clearly, a compact basis is desirable to limit the amount of computation required to produce the fitted curve. Depending on the form of $y(x)$, one set of functions may be compact in one case while another set may be compact in another case. This chapter will explore polynomials as basis functions.

11.2. POLYNOMIAL INTERPOLANTS

11.2.1. Introduction

An *interpolating function* is a function which, when evaluated at a given number of data points or control points, reproduces those points exactly and which approximates, or interpolates, along the pieces of the curve between the control points. The control points may arise from an actual function. In this case, the purpose is often to provide a simple function which interpolates a much more complex function over a limited domain. For instance, there may be a significant computational savings when a Gaussian function is interpolated with a low-degree polynomial over some domain of the independent variable. The control points may also arise from an experimental recording or from a simple attempt to draw a curve; in this case, there is no underlying function, and we merely desire the interpolating function to smoothly pass through the data points.

The theory of interpolating functions is extensive and well-developed. Early contributions were made by Newton and Lagrange, and most prominent mathematicians since have contributed to this field. Treatments can be found in most texts on numerical computing methods. The bulk of this theory is developed for polynomial interpolants, and we will explore this subject in some detail with a graphical approach. Because the nature of our study concerns the design and realization of curves and surfaces and not the accuracy of interpolation, we will bypass most of the formal theory related to accuracy.

FIGURE 11.2.1. Polynomial interpolation.

The requirement of the interpolating polynomial is that, given data points (x_k, y_k) with $0 \le k \le n$, the value of the polynomial satisfies

$$p(x_k) = y_k$$

for all k. It is important to note that there is only one polynomial of degree $\le n$ that satisfies the above for all the data points. That it may be less than n should be obvious when we take more than $n + 1$ control points exactly from a nth-degree polynomial. For instance, if we specify four points on a quadratic curve, then the second-degree, or quadratic, curve equal to the original can be made to fit them. Such cases are, of course, degenerate. As a corollary, we can state that except for degenerate cases a nth-degree polynomial is required to fit $n + 1$ data points.

This approach is a global fitting scheme because it uses all the points simultaneously. In the next chapter, we will look at the benefits of more local fitting schemes that use pieces of low-degree polynomials to fit, exactly or approximately, contiguous subsets of a larger set of numerous points calling for high-degree polynomials in the global scheme.

11.2.2.   Single-Valued Functions

Recall that functions are single-valued on a domain if, for every x value in the domain, at most one y value exists. This is the normal situation for polynomial interpolation. We will now see some examples that illustrate polynomial interpolants. In these examples, and others to follow, the data points will be shown as dots connected by straight lines and the interpolating polynomial will be shown as a simple continuous curve. The exact method of computing the interpolating polynomial can be found in almost any book on numerical analysis; basically, the $n + 1$ coefficients of the polynomial are determined exactly by the $n + 1$ data points. The plots in Figure 11.2.1 show examples of applying this method to 3, 7, and 11 data points.

These few examples immediately show the flaws in polynomial interpolants. Although the polynomial indeed passes through the data points, there can be large, and mostly undesirable, excursions between the data points. This is especially true at the two ends of the set of data points. This occurs in spite of the fact that the line segments connecting the data points themselves often appear to form a smooth curve. The problem increases with the number of data points, and wild fluctuations way beyond the desired range may occur. For second-degree or third-degree interpolating polynomials, it is seldom a problem though; this is why these low-degree polynomials

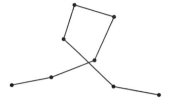

FIGURE 11.2.2. Data set with a loop.

FIGURE 11.2.3. Polynomial interpolation of the *x* points.

FIGURE 11.2.4. Polynomial interpolation of the *y* points.

FIGURE 11.2.5. Polynomial interpolation of the (*x*, *y*) points.

are popular for interpolating function values from those that are already tabled. (In such cases, a new interpolating polynomial is computed for each contiguous set of three or four tabled points.) In regard to curve design, almost certainly we would reject the polynomial fits in Figure 11.2.1 as being reasonable curves interpolating the control points. Clearly some other approach is needed.

11.2.3. Multivalued Functions

If a curve is multivalued on x as, for instance, with loops, the simple polynomial fit of y as a function of x cannot be used. However, the input points can be decomposed into a set of x points and a set of y points, both of which are single-valued in a parameter t. This parameter will be the cumulative length along the curve as measured simply by the straight lines connecting the points. We express this as

$$t_j = \sum_{i=1}^{j} [(x_i - x_{i-1})^2 + (y_i - y_{i-1})^2]^{1/2} \quad (0 \le j \le n)$$

It is convenient to set the first point at the origin; thus $(x_0, y_0) = (0, 0)$. Note that $t_0 = 0$. We now require both the (x, t) and (y, t) datasets to be exactly interpolated with an nth-degree polynomial as described in the previous section. Together these polynomials give the final interpolating function. An example will show how this works. We first plot in Figure 11.2.2 the (x, y) data points to be fitted. They are connected by a straight line to show the order of points; this example forms a simple loop. We then decompose these datapoints into the (x, t) and (y, t) datasets as described above. The plot of the (x, t) dataset and its interpolating polynomial is shown in Figure 11.2.3. Note the undesirable deviations of the computed curve from the simple straight-line connections near the endpoints. The plot of the (y, t) dataset and its interpolating polynomial is shown in Figure 11.2.4. Again note the problem near the endpoints.

FIGURE 11.3.1.   Interpolating polynomial of degree 4.

Finally, we plot the interpolated (x, y) curve in Figure 11.2.5 using the two interpolating polynomials generated by fitting the two parametric datasets. Note that the
curve passes through the data points exactly; however, this result is clearly a poor fit
to the intent of the data points which were meant to define a simple, smooth loop.
This example is not unusual in its failure to produce an acceptable result. As for the
single-valued function results, we conclude that interpolating polynomials will be of
little use in complex curve designs and that some other approach is needed.

11.3. POLYNOMIAL APPROXIMATIONS

11.3.1.   Approximation for Ordinary Curves

In contrast to interpolating functions that pass through the data points, we may construct approximating functions that only pass near the data points but are sufficiently
close to be acceptable. When we drop the requirement that the data points be exactly
matched, the approximating function will be usually much simpler and faster to compute than the interpolating function. In fact, approximation may be more desirable in
many cases where the experimental data points have measurement error or the points
input for curve design are not as smoothly placed as desired. With approximating
functions, we usually have more capability to produce desired effects in the shape of
the final function.

The possible polynomial approximating functions for a given set of data points is
limitless because they are no longer constrained to pass through the actual data points.
In practice, we may want to try fits with increasingly higher degree polynomials,
stopping when a suitable fit is obtained. The coefficients of the polynomial in this
case are computed by minimizing the sum of squares of deviations of the computed
curve from the data points. We can find expositions of this method in standard texts
on numerical analysis. Consider one of the examples in Section 11.2.2 which had 11
data points. The tenth-degree interpolating polynomial had unacceptable fluctuations
near the endpoints. We now fit this set of data points with a fourth-degree polynomial
as shown in Figure 11.3.1. Although this fit does not pass through the data points and
still appears to be inadequate, most observers probably consider it a better fit than the
interpolating polynomial.

11.3.2.   Approximation for Curves with Symmetry

The fitting function becomes simpler if we have a curve with symmetry, either even or
odd, about the y axis in mind. Recall that all even-degree monomials have symmetry
about the y axis and that all odd-degree monomials have antisymmetry about the y
axis. If the design goal involves symmetry, for instance, then the fitting polynomial
will have only even-degree terms.

FIGURE 11.3.2. Symmetric set of data points.

FIGURE 11.3.3. Polynomial approximation to x points.

FIGURE 11.3.4. Polynomial approximation to y points.

FIGURE 11.3.5. Polynomial approximation to (x, y) symmetric points.

Taking the data points that outlined a loop in Section 11.2.3, we can make this symmetric about the y axis by reflecting the last $n - 1$ of the n points around the first point. Then we can decompose this expanded set of $2n - 1$ points into x and y datasets as before, each fitting separately by a polynomial. Let $t = 0$ for the nth data point and let its x and y values be $(0, 0)$. The data points to be fitted now appear as shown in Figure 11.3.2. The (x, t) dataset is antisymmetric about the x axis; therefore, it is fit with a polynomial consisting of only odd powers of t. Figure 11.3.3 shows a fit to this dataset using a seventh-degree polynomial. (Lower-degree polynomials did not produce acceptable fits.) Note that there are actually 15 control points in this case; thus this is an approximation, not an interpolation.

Conversely, the (y, t) dataset is symmetric, so it is fit with a even polynomial. Figure 11.3.4 shows the sixth-degree polynomial fit. (Lower-degree polynomials did not produce acceptable fits.) Note again that this is an approximation, not an interpolation.

The test of our fit to the 15 data points will be to plot the resulting parametric curve made from the separate x and y fits and compare it with the 15 points. Figure 11.3.5 shows that the fit is rather good except at the end points. It must be emphasized that achieving this good fit requires relatively high-degree polynomials.

We have had some success with the simple example used here. Now let us apply this approach to a rather complex curve defined with the points in Figure 11.3.6. This is an antisymmetric set of 67 points.

First, we look at the polynomial approximation to the (x, t) points. The seventeenth-degree polynomial fit is shown in Figure 11.3.7. Lower-degree fits were inferior, and it still does not satisfactorily fit the data. The twentythird-degree polynomial fit for the (y, t) data is shown in Figure 11.3.8. (The y scale has been expanded to show the points more clearly.) This too is not a fully satisfactory fit; lower-degree fits were clearly poorer.

The final approximate curve, plotted from the polynomials fitting the x and y data, is shown in Figure 11.3.9. Although high-degree polynomials were used, this curve does not compare well with the shape outlined by the initial control points in Figure 11.3.6. Most observers would state that it fails to meet the design shape. Using higher-degree polynomials would not achieve a better result because large, unwanted fluctuations

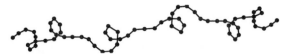

FIGURE 11.3.6. Antisymmetric set of data points.

FIGURE 11.3.7. Polynomial approximation to the *x* points.

FIGURE 11.3.8. Polynomial approximation to the *y* points.

FIGURE 11.3.9. Polynomial approximation to the (*x*, *y*) antisymmetric points.

would start to appear. Once again, we are led to conclude that polynomials are not a good basis set for fitting arbitrarily complex shapes.

11.4.   BEZIER CURVES

Named after a French automobile designer, *Bezier curves* are a special form of polynomial curve-fitting. Being able to largely overcome the problems seen in the previous sections, these curves are so widely used that they deserve expanded attention here. They are particularly noted for their smoothness: they do not try to pass through the data points themselves. However, the Bezier curve, as we will see, requires an nth-degree polynomial for $n + 1$ data points. It thus shares properties of polynomial approximation and polynomial interpolation.

We first introduce the *Bernstein polynomials* on which Bezier curves are based.

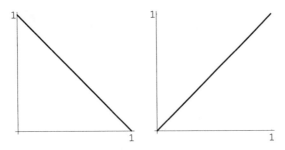

FIGURE 11.4.1. Bernstein polynomials for $n = 1$.

FIGURE 11.4.2. Bernstein polynomials for $n = 2$.

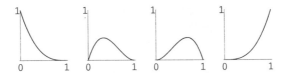

FIGURE 11.4.3. Bernstein polynomials for $n = 3$.

Let the parameter t be in the domain $[0, 1]$. These polynomials are defined on that domain by

$$B_{k,n}(t) = \frac{n!}{k!(n-k)!} t^k (1-t)^{n-k}$$

where $0 \le k \le n$. The coefficient here should be recognized as the binomial coefficient. As examples, we plot these polynomials for the cases of $n = 1, 2,$ and 3 in Figures 11.4.1, 11.4.2, and 11.4.3, respectively.

Now, let $n + 1$ data points be given by (x_k, y_k) with $0 \le k \le n$. We will call these the "control" points. Then the computed points of the Bezier approximating curve are given by weighting each Bernstein polynomial with its corresponding control point and summing thus:

$$x(t) = \sum_{k=0}^{n} x^k B_{k,n}(t)$$

$$y(t) = \sum_{k=0}^{n} y^k B_{k,n}(t)$$

The resulting curve blends the polynomials together. Each polynomial covers the entire range of the curve in the approximation but takes its largest value near the

FIGURE 11.4.4. Bezier curve approximations.

corresponding control point. In particular, note that the endpoints ($k = 0$ and $k = n$) will be fit exactly by the computed curve because only the first and last Bernstein polynomials are nonzero at these extremes. This is indicated by the graphical form of the Bernstein polynomials above; in general, this is true for any number of control points. We can thus form a closed curve by specifying the end control points as identical. In such a case, only continuity of the curve itself is guaranteed at this point. The curve will not smoothly pass through this point unless suitably constrained by nearby control points. Another property of Bezier curves at the endpoints is that the slope at either endpoint is equal to that of a simple line connecting that endpoint with the adjacent point.

The Bezier curve is defined on the parameter t. It can approximate curves that are multivalued in y versus x; that is, spirals, loops, and other similar features can be generated. However, as we shall see, some extra work is involved in making the Bezier curve satisfactorily approximate the control points in these cases.

Although the degree of the polynomials is n for $n + 1$ points, the amount of computation required to evaluate the Bernstein polynomial in the form given does not increase with n, provided that the binomial coefficients in the equation are computed and tabled prior to use. However, because any computed point of the Bezier curve requires a summation of $n + 1$ terms, the computation time required to produce one output point increases linearly with the number of input points.

An interesting property of the Bezier curves relates to their smoothness: the computed curve will lie within the convex polygon formed by the control points. Without regard to the number of control points, we draw the polygon which connects all or less than all of these points in a wholly convex manner. All control points will either be on this polygonal boundary or contained within it, and the Bezier curve itself will fall entirely within it.

Some examples of Bezier curves should clarify the manner in which they approximate control points. Although there is, in principle, no limit to the number of control points, the examples will use only a few. In Figure 11.4.4, the control points are plotted as large, connected dots; and the Bezier curve is evaluated on the domain $0 \leq t \leq 1$. All of these examples are open curves. Figure 11.4.5 shows an example in which the two endpoints are identical making a closed curve. The previous examples show that Bezier curves are quite smooth approximating functions. Often, the smoothing will be too much and the result will fail to satisfy the curve's design goals. There is a way to force the result to approximate the control points better: use multiple points at given locations. Using replicated points will require that additional Bernstein polynomials, with attraction to those points, be summed to produce the Bezier curve. The first and last point are excluded from this possibility because the Bezier curve already passes through them. The examples in Figure 11.4.6 show how

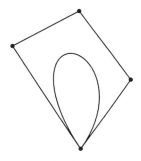

FIGURE 11.4.5. Bezier closed curve apporximation.

FIGURE 11.4.6. Bezier curve with middle control point used 1, 2, and 3 times .

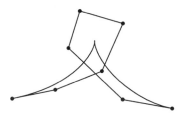 

FIGURE 11.4.7. Bezier curve fit to points defining a loop.   **FIGURE 11.4.8.** Bezier curve fit to points defining a loop, but with multiple points.

a quadratic Bezier curve approaches the second of three control points closer and closer as that point is used 1, 2, and 3 times.

Although the Bezier curve will often give a pleasing fit to the data points, it should be clear that construction of a Bezier curve to closely fit a large number of points attempting to define an even modestly complex shape will entail a high-degree polynomial and a large amount of computation. To achieve a Bezier fit to $n+1$ points, the highest degree of the required Bernstein polynomials is equal to n and may be significantly larger than the degree of an approximating polynomial discussed in Section 11.3.

Consider the set of control points used to define a shape having a loop in Figure 11.2.2. We already saw that the interpolating polynomials produced a poor fit to the data points and had extraneous bends in the curve. Let us now generate the Bezier curve fit to these points as shown in Figure 11.4.7. This too is a poor fit but for entirely different reasons. The Bezier curve is smooth, but it fails to produce the desired loop. By replicating points 4, 5, and 6 once, the Bezier curve can do better, as shown in Figure 11.4.8.

A better fit would be obtained for each additional replication of these points, but we

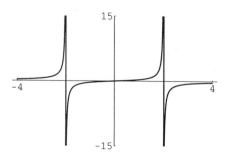

FIGURE 11.5.1.   Rational polynomial of degree 2: $x/(4 - x^2)$.

are already dealing with 10th-degree Bernstein polynomials (the original 8 points plus 3 replicated points). Thus continuous Bezier curves become less and less efficient as the fit is constrained to become better and better. Nevertheless, the result can almost always be made to look better than an ordinary polynomial approximation of any degree.

11.5. RATIONAL POLYNOMIALS

We introduced rational polynomials in Section 2.2. They are simply the ratio of two polynomials. We showed that the graph of a rational polynomial can be considerably more complex, for a given degree, than that of an ordinary polynomial. This suggests that this extra complexity can be utilized to successfully fit, in an approximate sense, sets of arbitrary data points difficult to fit with an ordinary polynomial.

We will illustrate the effectiveness of rational polynomials by considering a simple example. Start with the function $x/(4 - x^2)$, a rational polynomial having a highest degree of 2, plotted in Figure 11.5.1. Recall that the zeros of the denominator produce discontinuities in the graph of a rational polynomial: in this case at $x = -2$ and $x = 2$. Focus on the part of the graph between these points on the x axis where the function is continuous. We will attempt to fit this portion of the curve with an ordinary polynomial: The curve is sampled at the discrete x values of $-1.8, -1.6, -1.4, \ldots, 1.4, 1.6, 1.8$ for a total of 19 points to be approximated with the method of Section 11.3.1. The plots in Figures 11.5.2, 11.5.3, and 11.5.4 show the polynomial fit for degrees 3, 5, and 7, respectively. Note that the degree 3 and 5 polynomials show significant variations from the desired curve, but the degree 7 polynomial apparently gives a good fit.

If the same 19 points are used in a fitting algorithm employing the known form of the rational polynomial, then Figure 11.5.5 shows the result. The fit here is, of course, optimum because the known function has been used. The important point is that this known function, the rational polynomial, contains powers of x only up to two while the equally good fit using ordinary polynomials required a much higher power of x, at least equal to seven. In this case, we can see that the rational polynomial is a more compact basis than the ordinary polynomial basis and will be consequently more efficient to evaluate. However, other than for some very special arrangements of data points, the exact form of a rational polynomial may be difficult to determine. Practically, we may use ordinary polynomials even in cases where the data seems to require a rational polynomial.

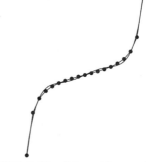

FIGURE 11.5.2. Polynomial approximation of degree 3 to middle section of $x/(4 - x^2)$.

FIGURE 11.5.3. Polynomial approximation of degree 5 to middle section of $x/(4 - x^2)$.

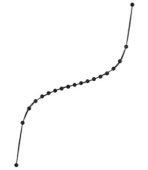

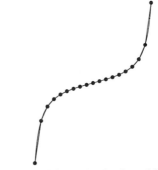

FIGURE 11.5.4. Polynomial approximation of degree 7 to middle section of $x/(4 - x^2)$.

FIGURE 11.5.5. Rational polynomial approximation.

11.6. SUMMARY

Polynomials are one type of basis functions used to find a curve that interpolates or approximates a given set of data points. Interpolation means the curve passes through all the data points exactly, and approximation means that it passes near them but not through them. Polynomials may perform well for interpolating a few data points, on the order of 3 or 4; but they rapidly deteriorate in their ability to globally interpolate sets of many data points. The fluctuations of the curve between data points becomes unreasonable and unacceptable for most purposes. In order to overcome this, we looked at approximating polynomials. These, too, produce poor fits to data sets having many points. In both cases, interpolation or approximation, the degree of the polynomial may be large and lead to inefficient computation of points on the curve.

Bezier curves were introduced as a means to overcome some of the poor fitting properties of ordinary polynomials. These curves are actually composed of polynomials of special form: the Bernstein polynomials. Bezier curves have some flexibility unavailable with ordinary polynomials and usually provide a more pleasing fit to a set of data points, especially when control is exercised by multiple use of certain data points in the solution. However, the Bezier forms introduced here are still a global solution to the approximation of data points and will require polynomials up to the nth degree for $n + 1$ data points and consequently will be inefficient to evaluate.

We examined the rational polynomial form as a means of fitting data points. It can give excellent results if the data points have certain arrangements amenable to the graph of such polynomials, but these arrangements are likely to be rare. The advantage of the rational polynomial, if it can be used, is that it will be compact and therefore efficient to evaluate.

Overall, the use of polynomials to fit data points generally leaves much to be desired. This is a result of the global nature of the approaches presented in this chapter. A more rewarding approach is to employ piecewise functions for fitting, which we will discuss in the next chapter.

Chapter 12

PIECEWISE CURVES

12.1. RATIONALE FOR PIECEWISE CURVES

The previous chapter showed that polynomials have undesirable features for fitting arbitrarily complex sets of (x, y) data. Part of the problem arises due to the fact that these functions provide a "global" fit to the data and must have a large number of terms. This chapter explores a different approach whereby the points can be fit more locally, resulting in a reduction in the number of terms required to realize the curve at any given point.

We will, in fact, find that cubic polynomials are usually adequate to fit any set of data points when the fitting is done locally. The resulting curves are actually composed of many small pieces that appear to merge together smoothly. For this reason, they are properly called *piecewise* curves.

The name *spline curve* is often associated with this type of fit because the result is very much like what hand drafting would produce when the flexible "spline" curve is used as a guide to draw a smooth curve through all the given points. Note, however, that the flexible spline is a continuous analog device. On the other hand, the piecewise approximations that we must compute digitally to obtain results similar to using a spline are discontinuous in the sense that different functions, or at least different parameters for similar functions, must be evaluated for the different pieces of the final curve.

12.2. CONTINUITY FOR CURVES

12.2.1. Definition of Continuity

Continuity is an important topic in algebra and geometry and certainly has clear implications for the appearance of curves. We make a formal definition of this property. Consider a function f defined on the domain D. This function is *continuous* at a point p_0 on D if, for some number $\varepsilon > 0$, there is a finite neighborhood U (contained in D) about p_0 such that $|f(p) - f(p_0)| < \varepsilon$ for every point in U. Essentially, we see a curve as continuous when it appears to be everywhere connected. We will label this continuity of the curve itself as "zero-order continuity," or "C^0 continuity".

In the case of parametric curves, the curve is graphed from two functions with the parametric variable of each defined on the same domain D. We can show that continuity of both functions is sufficient to guarantee continuity of the parametric curve itself and that the discontinuity of one or the other of the two functions is sufficient to make the parametric curve discontinuous.

Discontinuity of functions is often due to zeroes of the denominator of the expression for the function. Examples of discontinuous curves related to this are given in Figure 12.2.1. A zero in the expression for the denominator may not be sufficient to create a discontinuity, however. One of the most common functions exemplifying this is the sinc function, which is $\sin(x)/x$, introduced in Section 8.3. Its graph is

FIGURE 12.2.1. Discontinuity in the functions $1/x$, $x^2/(1-x^2)$, and $1/\sin(x)$.

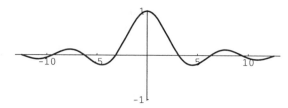

FIGURE 12.2.2. Removable discontinuity at $x = 0$ of sinc function.

shown again in Figure 12.2.2; note the continuity at $x = 0$. Such functions are said to have "removable discontinuities" and the functions plotted above have "essential discontinuities." It is not always easy to determine when a discontinuity is removable or essential. If you are interested, consult standard texts in advanced algebra for help in this regard.

12.2.2. Higher-Order Continuity

We can define higher-order continuity for curves from the derivatives. If the first derivative is continuous, then the curve is said to have C^1 continuity. Continuity of second and higher derivatives lead to C^2 and higher continuity. Curves with only C^0 continuity will have cusps or sharp corners, and curves with C^1 or higher continuity will appear to be smooth. The degree of smoothness depends on what derivatives are continuous. For practical purposes, we seldom require more than C^2 continuity to make a curve appear to be gracefully smooth. Curves with only C^0 continuity do have practical use in design though. Recall the cusps of the cycloid and trochoid in earlier chapters, for instance.

For polynomials, the maximum order of continuity is one less than the degree of the polynomial. Thus, everywhere, a straight line (degree = 1) has C^0 continuity, a quadratic curve (degree = 2) has C^1 continuity, and a cubic curve (degree = 3) has C^2 continuity. For harmonic functions, the derivatives of sine or cosine functions are infinite and are themselves sine or cosine functions; therefore, these functions have C-infinite continuity and are among the smoothest ones possible. The same can be said for the exponential function whose infinite derivatives are also exponential functions and will differ by, at most, a constant.

12.2.3. Piecewise Continuous Functions

This chapter treats piecewise curves, which are formed by joining end to end, two or more separate curve segments created from functions that differ in their coefficients. We will be concerned about the continuity of the composite curve as a whole.

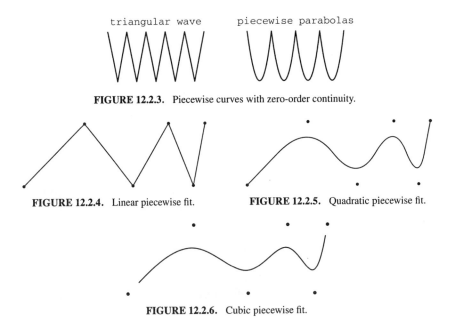

FIGURE 12.2.3. Piecewise curves with zero-order continuity.

FIGURE 12.2.4. Linear piecewise fit. **FIGURE 12.2.5.** Quadratic piecewise fit.

FIGURE 12.2.6. Cubic piecewise fit.

Piecewise curves without even C^0 continuity are not considered "curves" in the normal sense but are merely a collection of separate pieces. Piecewise curves with C^0 continuity are useful, however. Two examples of such curves are shown in Figure 12.2.3.

Most often though, we will want higher-order continuity for our piecewise curves; and we will first look at C^1 and C^2 continuity in a qualitative sense. Take a list of control points that define a triangular wave whose frequency increases linearly to the right as shown in Figure 12.2.4. The "curve" shown here may be described as a piecewise linear polynomial fit with five pieces; it has only C^0 continuity.

Our first attempt at fitting the given points, the linear fit, perfectly interpolates the points but leaves much to be desired as a curve. The next attempt in Figure 12.2.5 shows a piecewise quadratic fit to these control points (by a method yet to be described). There are actually four pieces to the curve as shown. The fit allows only C^1 continuity at the junctions between pieces. This curve appears to be fairly smooth, but carefully notice the junction positions roughly halfway between points 2 and 3, 3 and 4, and 4 and 5.

The last attempt in Figure 12.2.6 shows a piecewise cubic fit (by a method yet to be described). It appears to be smoother than the piecewise quadratic fit; this is due to the continuity of second derivatives in addition to first derivatives at the junction points.

The methods for attaining continuity of second or third derivatives at the points where the pieces of the curve join have been developed in numerous texts in the fields of numerical analysis and computer graphics. We will describe them in the following sections with only sufficient detail to understand the basics.

We must recognize that, while the true spline curve from hand drafting has continuity of derivatives to infinite order, certain compromises must be made in computing the

FIGURE 12.3.1. Bezier fit.

digital pieces: their junction points will lack continuity of higher-order derivatives. This is not really a problem in practice because the results do appear to be sufficiently smooth to the observer. This will be confirmed by the numerous examples in this chapter.

12.3. SPLINE APPROXIMATION

12.3.1. An Approach Based on Bezier Curves

Recall from Chapter 11 that a Bezier curve has the property that the tangent of the curve at either of the endpoints has a slope identical to that of a line connecting the first (or last) two control points. Thus, in constructing piecewise Bezier curves, we have an important tool that can be used to ensure continuity of the tangents of two adjoining curve segments of the piecewise curve. This provides the C^1 continuity.

Consider the set of seven control points in Figure 12.3.1 along with the sixth-degree Bezier curve which was fit to them using the method in Chapter 11. Note the tangency of the curve to the line joining the first and second points and again to that for the next-to-last and last points. As with many Bezier curves, this fit may not be acceptable; a closer tolerance to the control points is desirable in many circumstances.

This fit can be improved by duplicating points to bring the curve closer to the control points, as we saw in Chapter 11; but this is often a trial-and-error approach and leads to polynomials of higher and higher degree and consequently more computation. A piecewise fit is not only more efficient but will generally be more pleasing in appearance. Here is how we employ the principle governing the tangent of the endpoints to achieve C^1 continuity of the piecewise curve segments.

Let there be n original control points p_i with $1 \leq i \leq n$. Establish $n - 3$ new control points between points p_i and p_{i+1} for $2 \leq i \leq n - 2$. We place these new control points colinear with the surrounding pair. Their exact location between the pair is governed by the spacing of the control points. (Details can be found in Farin, 1988, Chapter 7.) These new points are the *junction points* for the $n - 2$ separate Bezier pieces. Each piece will be a quadratic Bezier fit to a triplet of adjacent points taken from the expanded set of control points. We will use the last point of one triplet as the first point of the next triplet. Therefore, the tangents of the ends of adjacent curve pieces will be identical and will be equal to the that of the line connecting the two original control points. The mathematical details are omitted here, but the C^1-continuity piecewise curve is shown in Figure 12.3.2. The original and the added control points are shown also.

FIGURE 12.3.2. Quadratic Bezier piecewise fit.

The piecewise curve provides a good fit to the original control points and passes exactly through the added points. The junction points are clearly represented on the piecewise curve: they are identical with the added control points. We should also mention another set of one-dimensional points associated with the piecewise curve. Consider the final curve as a parametric function with parameter t. As t varies from 0 at the start of the curve to 1 at the end of the curve, those values of t at which a new quadratic parameterization begins are termed the *knots* or *breakpoints*. Note that knots are associated with the parameter t and junction points are associated with actual (x, y) values. There is, of course, a one-to-one mapping from the knots to the junction points in this case.

We can extend this approach to piecewise cubic curves in order to achieve C^2 continuity. The essence of this extension is to add new control points so that they are colinear with two pairs of adjacent points used for the piecewise quadratic curves. (Details can again be found in Farin, 1988, Chapter 7.)

While piecewise Bezier curves share many of the properties of single Bezier curves, most notably the convex-hull property and the fact that they pass through the end control points, they have several other beneficial features. Piecewise curves are more locally controlled because no one control point will affect a piece beyond either of its nearest junction points. Thus, movement of a control point will have very local and predictable results for piecewise curves, and similar movement will cause some change to the whole of the single Bezier curve fitting the same set of control points. Piecewise curves, often consisting of quadratic or cubic pieces in practice, are much faster to evaluate than single Bezier curves, although there is somewhat higher overhead due to the necessity of keeping track of knots and separate curve segments.

12.3.2. Periodic Splines—Open Curves

Although the procedure outlined and illustrated in the previous section is adequate for constructing piecewise curves, it is somewhat cumbersome and is really difficult to extend to higher-degree polynomials than cubic. Also, the control points are not "blended" into the final curve in a uniform manner with that procedure; however, this fact is usually of little consequence in the final appearance of the curve, except near the endpoints.

We seek a general algorithm by which spline curves can be easily constructed. There are numerous treatises on the computation of spline curves, and most of them formulate the solution as a blending of the control points into the final curve. The blending functions form a "basis", in the sense of Section 11.1, for the final piecewise curve; and these basis functions are weighted by the actual values of the points. When these basis functions have the minimum possible length for the particular degree of polynomial used, the spline is known as a *B-spline* ("B" for basis).

In this section, we outline a procedure for piecewise *B*-splines which is due to Mortenson (1985). It does blend the control points in a uniform manner and is, additionally, quite easy to code into a program for constucting *B*-splines.

Let $k - 1$ be the degree of the polynomials used to fit the control points piecewise. The individual control points are again given as p_i with $1 \leq i \leq n$. Let the individual curve segments be given by $s_i(u)$ with $1 \leq i \leq n - k + 1$ and $0 \leq u \leq 1$. There will be $k - 1$ fewer curve segments than control points. For instance, when $k = 2$ (linear fit), there will be one curve segment for each pair of adjacent control points (a straight line connecting them). The parameter u varies over $[0, 1]$ for each curve segment. Mortenson (1985) derives the following compact expression for piecewise curves which have degree $k - 1$:

$$s_i(u) = \mathbf{U}_k \mathbf{M}_k \mathbf{P}_k$$

where

$$\mathbf{U}_k = [u^{k-1} u^{k-2} \ldots u^1],$$
$$\mathbf{M}_k = k\text{-by-}k \text{ matrix of constant coefficients,}$$
$$\mathbf{P}_k = [p_i \, p_{i+1} \ldots p_{i+k-1}]^T$$

The superscript T indicates transpose. The x and y components of s are computed separately using the x and y values of the p_i. In general, any particular p_i appears in k segments of the piecewise curve. It is blended into these k segments by the vector $\mathbf{U}_k \mathbf{M}_k$ (the blending function) which is invariant in form. Thus, each point is blended into the final curve in an identical manner; this fact leads to calling the result a "periodic *B*-spline." Thus, we see the blending of the points to be very localized. The $\mathbf{M}_k$ matrices for $k = 1, 2, 3, 4$ are

$$\mathbf{M}_1 = 1$$

$$\mathbf{M}_2 = \begin{bmatrix} -1 & 1 \\ 1 & 0 \end{bmatrix}$$

$$\mathbf{M}_3 = \frac{1}{2} \begin{bmatrix} 1 & -2 & 1 \\ -2 & 2 & 0 \\ 1 & 1 & 0 \end{bmatrix}$$

$$\mathbf{M}_4 = \frac{1}{6} \begin{bmatrix} -1 & 3 & -3 & 1 \\ 3 & -6 & 3 & 0 \\ -3 & 0 & 3 & 0 \\ 1 & 4 & 1 & 0 \end{bmatrix}$$

$\mathbf{M}_1$ merely reproduces the control points and is trivial. $\mathbf{M}_2$ connects the control points with straight lines. $\mathbf{M}_3$ produces a quadratic *B*-spline with C^1 continuity. $\mathbf{M}_4$ produces a cubic *B*-spline with C^2 continuity. In the case of $\mathbf{M}_3$ and higher-degree fits, the piecewise curve does not pass through the endpoints but stops somewhere between the endpoints and their adjacent points. This is perhaps an undesirable feature of the periodic *B*-spline, but it can be overcome, to some extent, as discussed below.

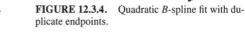

FIGURE 12.3.3. Quadratic *B*-spline fit.

FIGURE 12.3.4. Quadratic *B*-spline fit with duplicate endpoints.

FIGURE 12.3.5. Cubic *B*-spline fit.

FIGURE 12.3.6. Cubic *B*-spline fit with duplicate endpoints.

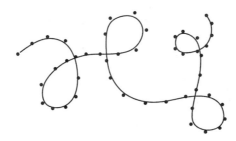

FIGURE 12.3.7. Cubic *B*-spline fit to 48 points.

Consider the control points used in Section 12.3.1 again. We apply the periodic *B*-spline algorithm to them with quadratic ($k = 3$) fitting; the result is shown in Figure 12.3.3. The curve does not extend to the endpoints but actually stops roughly midway between the first and second control points and again between the last two points. However, this can be alleviated by simply duplicating the two end control points. The result of recomputing the curve with the duplicate endpoints is shown in Figure 12.3.4. It is not perceptibly changed from that in Figure 12.3.3 but now extends to the endpoints.

Figure 12.3.5 shows the cubic ($k = 4$) fit to the same control points using the periodic *B*-spline algorithm. The cubic *B*-spline, just as the quadratic one, does not extend to the endpoints. This is overcome by again duplicating the first and last control point and then recomputing the curve; the result is shown in Figure 12.3.6.

To end this section, we show an example of how well a piecewise cubic *B*-spline will fit a contorted sequence of numerous control points. The 48 data points and the cubic *B*-spline fit are shown in Figure 12.3.7. The result is very nearly what we might produce if asked to draw a freehand curve through the points.

12.3.3. Periodic Splines—Closed Curves
To produce closed curves with the method of Section 12.3.2, it is not necessary to make the endpoints coincident as for ordinary Bezier curves but only to modify the computation procedure somewhat. There will be as many curve segments as there are control points, regardless of the order of continuity imposed, because we can use

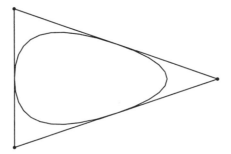

FIGURE 12.3.8. 3-point quadratic B-spline closed-curve fit.

the control points in a "circular" fashion. We can pick up points from the end of the curve when we compute the start of it and conversely pick up points from the start of the curve when we compute the end. For instance, in the quadratic-fit case, the first curve segment will be given by

$$s_1(u) = \mathbf{U}_3 \mathbf{M}_3 [p_n p_1 p_2]^T$$

In the cubic-fit case, the first curve segment is

$$s_1(u) = \mathbf{U}_4 \mathbf{M}_4 [p_n p_1 p_2 p_3]^T$$

Similarly, at the end, the p vector is wrapped around to the first point for the quadratic-fit case

$$s_n(u) = \mathbf{U}_3 \mathbf{M}_3 [p_{n-1} p_n p_1]^T$$

and to the first and second points for the cubic-fit case

$$s_n(u) = \mathbf{U}_4 \mathbf{M}_4 [p_{n-1} p_n p_1 p_2]^T$$

There is, in fact, no preference for start and endpoints in this algorithm. The curve segments can be computed starting with any set of 3 (quadratic) or 4 (cubic) adjacent points. This algorithm is therefore somewhat simpler than that for the open-curve B-spline fit. We start with the list of control points and merely rotate them by one to compute each new curve segment. When the list has been rotated back to the original order, the B-spline curve has been completed. Some examples of closed curves made with the periodic B-spline algorithm are shown in Figures 12.3.8 to 12.3.10.

The closed, periodic, B-spline algorithm for cubic fitting is next applied to the points of Figures 12.3.9 and 12.3.10. (The example from Figure 12.3.8 has one less point than needed for cubic fitting.) The results are shown in Figure 12.3.11 and 12.3.12. We can see that the cubic fits, in comparison with the quadratic fits, sacrifice proximity to the control points to produce a smoother curve.

The subject of B-splines and related splines is large, fertile, and continuously growing. Many modifications and improvements have been suggested. This section has only discussed, in detail, one practical approach (periodic B-splines) that should serve well in a wide variety of contexts for which approximation, not interpolation, is sufficient. Further study is warranted if you feel that B-splines will be regularly used in your graphics design.

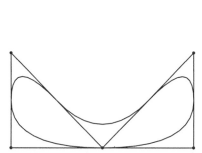

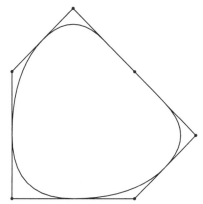

FIGURE 12.3.9. 5-point quadratic *B*-spline closed-curve fit.

FIGURE 12.3.10. 6-point quadratic *B*-spline closed-curve fit.

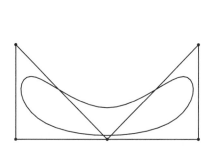

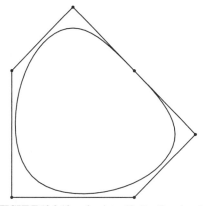

FIGURE 12.3.11. 5-point cubic *B*-spline closed-curve fit.

FIGURE 12.3.12. 6-point cubic *B*-spline closed-curve fit.

12.4. SPLINE INTERPOLATION

12.4.1. Open Curves

Often it is not appropriate or sufficient to approximate the control points, and we require the fitted piecewise curve to interpolate, or pass through, the control points. Clearly, this could be done with straight lines connecting the points; but this provides only C^0 continuity and is almost never acceptable. We will present a method, described in Farin (1988, Ch. 9), that will use interpolating cubic curve segments and provide C^2 continuity so that the result appears to be smooth. The method establishes two new control points between all original points in such a manner that the cubic polynomial fit to each set of two original points plus the two new points will produce curves which are C^2 continuous at the knots. The result is usually referred to as a "piecewise interpolating cubic spline." We compute the cubic spline by solving a tridiagonal set of linear equations. In setting up the equations that compute the curve, we can impose a variety of "end conditions" to produce desired effects at the

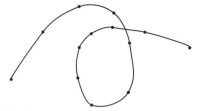

FIGURE 12.4.1. Cubic spline interpolation for simple curve.

FIGURE 12.4.2. Cubic spline interpolation for a loop.

endpoints. The method is applied here with "natural" end conditions. These force the curve to have a tangent at the endpoints equal to that of the line connecting the endpoints with their respective adjacent points; this usually provides a satisfactory and pleasing appearance to the result. Recall that this is a property of Bezier curves.

Taking the set of seven control points from the curve which has been approximated with several methods already, we compute the interpolating spline and plot it with the original points in Figure 12.4.1. Note that it does interpolate all the points and has a smooth appearance also. This should mimic well the piecewise spline curve drawn by a person through these same points. In fact, a person's attempt to "connect the dots" should be similar to this piecewise cubic spline fit. Such interpolating splines can handle any sequence of control points. The next example in Figure 12.4.2 shows the result when the control points specify a loop.

12.4.2. Closed Curves

If we desire the interpolating curve to close on itself, we must take an approach such as in Section 12.3.3 and use the control points in a circular, or rotating, sense. There will be no "end conditions" as in open curves, and the piecewise curve is derived in a uniform manner for all segments. However, this leads to a linear system of equations that requires some additional computation to solve.

Let us look first at how the cubic spline will interpolate four points arranged on the corners of a unit square. The result, as shown in Figure 12.4.3, is a smooth, continuous curve identical in each of its four segments between the knots. The result for the square leads us to predict that the result of interpolating the corners of polygons with larger and larger number of edges will produce closer and closer approximations to a circle. This is confirmed by interpolating the corners of a decagon (10-sided polygon), as shown in Figure 12.4.4.

Loops and other complexities are treated well by the cubic spline interpolation algorithm. In Figure 12.4.5, we fit the control points of a figure "8". When the points of a closed curve are to be fitted exactly, the cubic spline interpolation is the means of most practical use. If the appearance of such a piecewise curve is not quite what we want, then adding control points at desired locations should readily bring our result into compliance with the design goal.

Interpolating splines produce excellent smooth curves while satisfying the requirement that the curve pass through all the control points. The result, in most cases, should be very close to what we would produce by manually drawing a line through the given points.

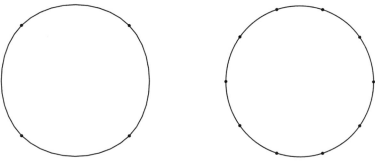

FIGURE 12.4.3. Cubic spline interpolation for a square.

FIGURE 12.4.4. Cubic spline interpolation for a decagon.

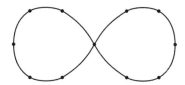

FIGURE 12.4.5. Cubic spline interpolation for a figure eight.

12.5. SUMMARY

Curves that are computed piecewise to interpolate or approximate a set of control points will behave well and produce results similar to manual procedures. The computation is less than that required for global curve methods, which often do not produce a satisfactory result. The order of the continuity of the pieces is important for the overall appearance of the curve, and continuity of second derivatives is preferred in almost all cases. Various algorithms exist for producing piecewise curves which approximate a set of control points. An approach based on Bezier curves was presented as well as one based on periodic splines. Both open and closed curves can be generated with these methods. Spline methods for interpolation of control points were also presented; these too can handle open and closed curves. Spline curve methods, in general, allow for rapid curve recalculation to fit new or translated control points due to the fact that the curve pieces are very locally controlled.

Chapter 13

COMPLEX CURVES FROM HARMONICS

13.1. INTRODUCTION TO FOURIER SERIES

Although most parts of this chapter would traditionally be approached through the theory of Fourier integrals or transforms, this approach does entail mathematical abstractions inappropriate for the general reader. They include integrals and the complex numbers, which are topics beyond what is considered requisite knowledge for this book. What is planned for this chapter can be accomplished through the less rigorous application of Fourier series even though, in some exceptional cases, this has drawbacks.

Previously, in Chapters 5 and 6, many examples were given of curves that were the sum of two or more harmonic functions. The fundamental period was introduced as an important concept, and it is again defined as that length L such that

$$g(x + L) = g(x)$$

for a function g defined on x. Thus, because the function repeats infinitely with a period L, we can generate the entire function solely from one piece of length L by an infinite number of translations.

We now put forth an extremely important theorem due to the French scientist Joseph Fourier. It is that (practically) any function can be expressed as an infinite sum of harmonics. For a function $g(x)$, the theorem would read

$$g(x) = a_0 + \sum_{n=1}^{\infty} [a_n \cos(2\pi n f_1 x) + b_n \sin(2\pi n f_1 x)] \quad (0 \le x \le L)$$

Here f_1 is the fundamental frequency, the inverse of the fundamental period L. It should be recognized that the fundamental period L may be infinite (for instance, the function $\exp(-ax^2)$ has an infinite period). In such cases, f_1 becomes infinitesimal. As we see in the above summation, the harmonics (sine and cosine) are a set of basis functions discussed in Section 11.1. They may or may not be compact, depending on the form of $g(x)$.

In the reverse direction, given $g(x)$, we can find the coefficients a_n and b_n by integration of the function, weighted by the corresponding sine or cosine function, over the fundamental period:

$$a_n = \frac{2}{L} \int_{-L/2}^{L/2} g(x) \cos(2\pi n x / L) dx \quad (n = 1, 2, 3, \ldots)$$

$$b_n = \frac{2}{L} \int_{-L/2}^{L/2} g(x) \sin(2\pi n x / L) dx \quad (n = 1, 2, 3, \ldots)$$

The special value a_0 is computed separately by

$$a_0 = \frac{1}{L} \int_{-L/2}^{L/2} g(x) \, dx$$

205

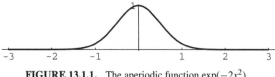

FIGURE 13.1.1. The aperiodic function $\exp(-2x^2)$.

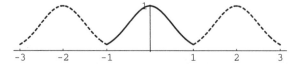

FIGURE 13.1.2. The function $\exp(-2x^2)$ folded into a periodic function.

(We have used integrals here, inconsistent with our overall intent; if you are unfamiliar with them; they should be seen as just special forms of summation.) We note that a_0 is the mean value of the function. These expressions, involving integrals, are not suitable for general curves. For this reason and because, in the forward direction, the series is summed over an infinite number of frequencies, the above formulas are not practical for use.

A formula that can be used successfully in most cases of practical interest is the finite Fourier series:

$$g(k\Delta x) = A_0 + 2 \sum_{n=1}^{N/2-1} [A_n \cos(2\pi nk/N) + B_n \sin(2\pi nk/N)]$$
$$+ A_{N/2} \cos(\pi k) \quad (1 \le k \le N)$$

This is our *Fourier synthesis* formula and will be the main tool of this chapter. It constructs a "sampled" function of length $(N-1)\Delta x$ with Δx being the sampling increment. We assume that the length is to be the fundamental period L. In the case of functions with infinite period, some "significant" portion of the function is used; and we assume that this portion takes on the characteristic of a function which is periodic with the length L of this portion. Thus, in these cases, the function that is represented by the series here is not, in fact, the original function.

Consider the function $\exp(-2x^2)$, known to have infinite period; this is plotted in Figure 13.1.1 for $-3 \le x \le 3$. If we apply the finite Fourier series to it, then it must be truncated to some length L, say 2, as shown in Figure 13.1.2 with a solid line for $-1 \le x \le 1$. The assumed part of the represented function is shown as dotted-line portions, which actually repeat to infinity on either side. The finite Fourier series represents the entire graph shown, not just the solid line. It should be recognized that the Fourier series representation of the truncated and repeated curve will not match the original function except in the interval from -1 to $+1$. The Fourier series representation of infinite functions like the one shown here can be improved by increasing the interval L, at the expense of having proportionately more Fourier terms.

A further comment on the finite Fourier series is that it is a complete interpolating representation of the sampled function $f(x)$ because it will pass through the sampled points exactly. It knows nothing, however, about the points between sampled points.

Given the sampled function $g(k\Delta x)$, the Fourier coefficients are calculated as follows:

$$A_n = \frac{1}{N} \sum_{k=1}^{N} g(k\Delta x) \cos(2\pi n k/N) \quad (0 \le n \le N/2)$$

$$B_n = \frac{1}{N} \sum_{k=1}^{N} g(k\Delta x) \sin(2\pi n k/N) \quad (0 \le n \le N/2)$$

These are our *Fourier analysis* formulas. It should be noted, from the properties of sine and cosine functions, that there are these special relations:

$$A_0 = \frac{1}{N} \sum_{k=1}^{N} g(k\Delta x)$$

$$A_{N/2} = \frac{1}{N} \left[\sum_{k=2}^{N,2} g(k\Delta x) - \sum_{k=1}^{N-1,2} g(k\Delta x) \right]$$

$$B_0 = B_{N/2} = 0$$

The notation "$N, 2$" means summation to N by every other 2; likewise "$N - 1, 2$" means summation to $N - 1$ by every other 2. The Fourier coefficient A_0 is equal to the average of the function g over the interval of N sample points. The coefficient $A_{N/2}$ is equal to the difference of the averages of the odd-numbered samples and the even-numbered samples. The fact that $B_0 = B_{N/2} = 0$ means that 2 of the $N + 2$ coefficients defined by the relations above provide no information on N samples of the function g. In effect, there are only N coefficients which matches the number of samples of g exactly; therefore, it neither increases nor decreases the information of the sampled function g.

The coefficients A_n and B_n of the Fourier series give the amplitude of the respective cosine and sine waves which comprise a function. For any n, A_n and B_n represent the same frequency of the wave, only differing in the phase. Therefore, it is common to combine them vectorially and state that the amplitude of a particular frequency component associated with n is

$$C_n = [A_n^2 + B_n^2]^{1/2} \quad (0 \le n \le N/2)$$

The plot of C_n versus frequency nf_1 is referred to as the *amplitude spectrum*, or simply spectrum, of the function. The phase in radians of the spectral component at frequency nf_1 is given by

$$\phi_n = \arctan(B_n/A_n)$$

When $B_n = 0$, the phase is zero, which is appropriate to a cosine wave. When $A_n = 0$, the phase is $\pi/2$, which is appropriate to a sine wave.

Another important property of the finite Fourier series representation of a function is that it is an "orthogonal" basis of the function in the interval 0 to L. Taking any sine and cosine functions of the series to be u and v, it can be shown that

$$\sum_{k=1}^{N} u(2\pi nk/N)v(2\pi mk/N) = 0$$

unless u and v are either both sin or both cos and $n = m$. In that case, the result is $N/2$ (except for the special value $n = m = 0$ where the result is N for u and v both cos and the result is 0 for u and v both sin).

In the following sections of this chapter, we will consider all curves to be segments of infinite curves and have length L and so apply the finite Fourier series as given above. In computing practice, this is the only choice because we cannot deal with infinite regions for computation. Within this computing context, the Fourier series representation is usually implemented by the *discrete Fourier transform* which is a computationally efficient and accurate way of computing the Fourier coefficients given a function and, conversely, computing the function given the Fourier coefficients. Details of this discrete transform are not given here because it requires that the discussion lead into complex numbers. If you are interested, you can find descriptions of the discrete Fourier transform in most books on time series analysis. Here we will use the computationally less efficient Fourier series given earlier in this section.

13.2. FOURIER ANALYSIS AND SYNTHESIS FOR CURVES

13.2.1. Symmetry and Antisymmetry

Previously, we have examined symmetry of curves in several chapters. Symmetry relations for a function have some simple counterparts in the Fourier domain. First consider that the Fourier domain is composed of sine and cosine functions. The former is antisymmetric and the latter is symmetric, regardless of the frequency of the function. They are, therefore, exclusive. If a function is symmetric, it will contain only cosine terms in the Fourier representation ($B_n = 0$ for all n). If it is antisymmetric, the Fourier representation must consist of sine functions only ($A_n = 0$ for all n). The converse of these statements must also be true. Therefore, any function, if expressible in a Fourier series, must be a combination of a purely symmetric and a purely antisymmetric part. In Chapter 2 we explained that symmetric functions could be changed to antisymmetric ones and vice-versa through multiplication or division by monomials. For instance, the symmetric function $x^2 - 1$ becomes antisymmetric when multiplied by x as shown in Figure 13.2.1. Does this alter the Fourier representation beyond changing from cosine to sine functions? The Fourier amplitudes C_n for both functions are shown in Figure 13.2.2 when the interval of analysis is set to $-1 \leq x \leq 1$ and $\Delta x = 0.04$. For these functions, $C_n = |A_n|$ and $C_n = |B_n|$, respectively. Note that the logarithm ($\log C_n$) of the Fourier spectrum is plotted.

By comparing these amplitude spectra, we see that the function $x^2 - 1$ apparently has high-frequency components with larger amplitude than does $x(x^2 - 1)$. This must be seen in the context of the truncation of the two functions at -1 and $+1$, however. The truncation of the later is at the points where, if the function were repeated, there is neither a discontinuity of the function itself nor of its first derivative. This causes it to appear smooth to the Fourier series representation, meaning small amplitude at high frequencies.

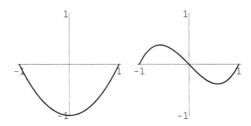

FIGURE 13.2.1. Graph of the functions $x^2 - 1$ and $x(x^2 - 1)$.

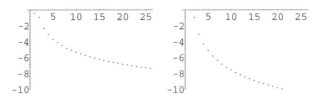

FIGURE 13.2.2. Logarithm(Fourier amplitude) for $x^2 - 1$ and $x(x^2 - 1)$.

If the coefficients A_n of the first function, $x^2 - 1$, are merely switched to the sine series, the sine series will have an equivalent spectrum. There is just one problem: the coefficient for the zero-frequency cosine wave. Clearly, this cannot be used in the sine series and must be discarded. If this sine series is then inverted to produce a curve as shown in Figure 13.2.3, it will be antisymmetric but not equivalent to the graph of $x(x^2 - 1)$ in Figure 13.2.1. There is, as expected, more high-frequency character.

We have shown that an antisymmetric curve can be created from a symmetric one while preserving the frequency content via the Fourier series representation. Similarly, the converse can be accomplished. This only requires that the coefficients of the Fourier terms be switched from a pure cosine series to a pure sine series. In accomplishing the converse fron sine to cosine terms, the value of the zero-frequency cosine coefficient can be set arbitrarily. This merely affects the vertical position of the resulting curve, not its shape.

13.2.2. Discontinuities at Ends of the Curve
Taking a given curve portion, imagine it to be wrapped around so that the end at $-L/2$ is placed next to that for $L/2$. If the curve for which a Fourier series representation is sought has a vertical discontinuity at these endpoints, the spectrum of the curve is strongly affected at the high frequencies because significant high-frequency components are required to match the jump at the discontinuity. Recall that in Section 13.1 the Fourier series is really approximating the curve as if it repeated infinitely outside the actual range of the analysis. An example will help to illustrate the effect. Consider the cubic curve $x(x^2 - 1)$ used above. It is plotted for $-1.5 < x < 1.5$ in Figure 13.2.4. The zeroes of the function are at -1 and $+1$.

The portion between -1 and $+1$ on the x axis was analyzed in the previous section to get the spectrum indicative of a smooth function. If the range of analysis is extended, say from -1.2 to 1.2, then the ends of the function no longer meet when

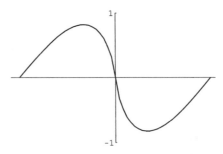

FIGURE 13.2.3.   Inversion of the swap of cosine for sine series.

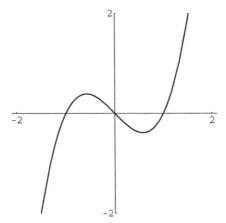

FIGURE 13.2.4.   Graph of the function $x(x^2 - 1)$.

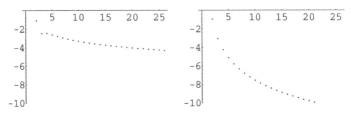

FIGURE 13.2.5.   Logarithm(Fourier amplitude) for domains of $[-1.2, 1.2]$ and $[-1, 1]$.

that section is repeated. The discontinuity is exactly $g(-1.2) - g(1.2)$. This "step" must be represented within the calculated Fourier series shown in Figure 13.2.5 next to that for the original series with $-1 < x < 1$. Note that the curve portion with the discontinuity has high-frequency amplitudes that are several logarithm units greater than the corresponding spectral amplitudes for the portion without discontinuous ends.

When we analyze a symmetric function, however, the window $[a, b]$ should always be chosen so that it falls evenly about the point of symmetry. This will ensure that $g(a) = g(b)$ and that the endpoints are continuous (at least C^0 continuity). The spectrum will consequently have small amplitudes at the highest frequencies, provided

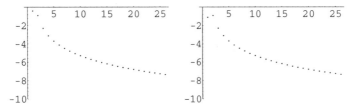

FIGURE 13.2.6. Logarithm(Fourier amplitude) for $x^2 - 1$ and x^2.

that there are no discontinuities internal to the interval $[a, b]$. At the other end of the spectrum, there is a strong effect due to the fact that the zero-frequency cosine coefficient is the average of the function over the interval $[a, b]$. In rare cases, this average is equal to zero. By adding a constant to the curve, this zero-frequency coefficient is increased or decreased in proportion to the change of constant. Consider the function $x^2 - 1$ used in the previous section. The mean of this function over the interval $[-1, 1]$ is exactly $-2/3$, and the Fourier series starts with a coefficient equal to this. If we add unity to this function to get just x^2, the mean becomes $+1/3$. The spectrum (log C_n) of both functions is shown in Figure 13.2.6. The first (zero-frequency) coefficient has changed, as expected, due to the difference in mean between the two functions ($-2/3$ for the first and $1/3$ for the second). The rest of the two spectra are identical.

If we add a constant, say c, to an antisymmetric curve, for which the spectrum consists entirely of sine waves, the effect of the constant will appear in a zero-frequency cosine term; but no other cosine terms will become nonzero. In this case, the combined function has its symmetric part equal to a straight horizontal line at $y = c$.

13.2.3. Synthesis of Curves with Discontinuities
It is often useful to employ Fourier series, which consist of smooth harmonic functions, to approximate functions which are not themselves considered to be smooth in any sense. Such curves may have internal discontinuities of zero-order or first-order.

Consider the *step function* $g(x)$ which is equal to -1 for $x < 0$ and $+1$ for $x > 0$. There is a zero-order discontinuity at $x = 0$. This function is antisymmetric and will, therefore, be represented by a sine series only. Set the analysis window from -1 to $+1$ and use $\Delta x = .04$ and compute the sine coefficients ($1 \leq n \leq 24$) using the Fourier analysis formula from Section 13.1. The full series of 24 coefficients will exactly reproduce the function at the sampled points. What if we use less than 24 in the summation to produce the function g? The series of plots in Figure 13.2.7 shows the reconstruction of the step function using the first 5, then the first 15, and finally all 24 sine-wave components. The approximation improves with the number of coefficients. Note that the function is reconstructed with 24 components as though it had a fundamental period of $L = 2$ with a discontinuity every unit in x.

Another similar function is the triangular wave shown in Figure 13.2.8. It has first-order discontinuities at the top and bottom of each straight-line segment. Take the analysis window to again be -1 to $+1$ with $\Delta x = .04$ and compute the Fourier cosine series of 26 terms ($0 \leq n \leq 25$). (Recall that the first term, for $n = 0$, is the

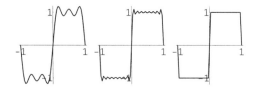

FIGURE 13.2.7. Approximation of step function with 5, 15, and 24 sine waves.

FIGURE 13.2.8. Triangular wave.

FIGURE 13.2.9. Approximation of one cycle of a triangle wave with 3, 6, and 26 cosine waves.

average of the function over the analysis interval.) Then reconstruct the function with the first 3 terms, the first 6 terms, and finally all 26 terms as shown in Figure 13.2.9.

By using the full finite series, an original function can be reproduced; but, by using only a few of the terms of the series, approximations to the original function can be made. These approximations, which can be computed rapidly, may be useful in many cases and may provide curves that are of interest for design purposes. We have shown examples using curves with discontinuities of zero or first order, but the method can be applied to smooth, continuous curves having algebraic or transcendental terms or factors.

13.2.4.   Synthesis of Curves by Random-Amplitude Fourier Series

We have shown that a curve can be synthesized from a Fourier series. Thus far, such curves have had shapes that correspond to, or closely follow, some mathematical function. It is possible to synthesize curves which are not associated with any such functions, and their analytic behavior is solely defined by the Fourier series from which they are formed.

In this context, we will take a random approach to the generation of curves by randomizing the coefficients of the harmonic components. Those curves of most interest are symmetric or antisymmetric; so we will randomize the coefficients of either the cosine terms (A_n) or the sine terms (B_n) only. The Fourier synthesis formula in Section 13.1 calls for $N/2$ sine terms and $N/2$ cosine terms for a discrete function of length N. We can, however, use less than $N/2$ such terms assuming that the remainder have coefficients equal to zero. In the following, we will experiment with such "truncated" Fourier synthesis formulas.

The set of curves in Figure 13.2.10 shows six realizations of randomizing the first six coefficients of the cosine series (not including A_0) and then synthesizing the

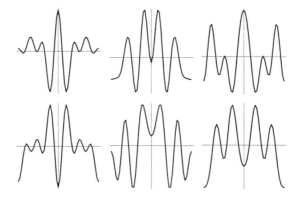

FIGURE 13.2.10. Functions created from 6 random-amplitude cosine terms.

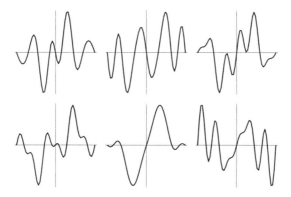

FIGURE 13.2.11. Functions created from 6 random-amplitude sine terms.

curve from these six random-amplitude components. The amplitude values were determined by randomly selecting a value between $-1/2$ and $+1/2$. The curves were synthesized at 51 points from -1 to $+1$ and then normalized to have a maximum amplitude of unity prior to plotting. The next set of curves in Figure 13.2.11 shows similar results for six random-amplitude sine components. A series based on random sine terms can be summed with another series based on similar cosine terms to create a curve without any symmetry. Figure 13.2.12 shows six realizations of this procedure.

By randomly generating curves in this manner, it is inevitably that shapes will appear that are not naturally conceived and generated. As done here, these curves are best described as "random noise." Some deterministic influence can be added by perhaps weighting the random coefficients in the sine and cosine series with a factor that depended on the frequency. In this way, high-frequency behavior or low-frequency behavior could be emphasized.

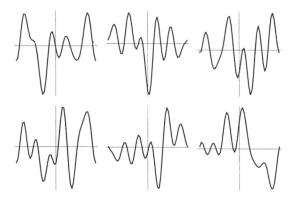

FIGURE 13.2.12.   Functions created from 6 random-amplitude sine terms and 6 such cosine terms.

13.2.5.   Synthesis of Complex Parametric Curves

The parametric form for plane curves

$$x = f(t); \quad y = g(t)$$

was shown in previous chapters to enable the generation of very complex curves even when the functions $f(t)$ and $g(t)$ are themselves relatively simple. Examples with no more than two terms in these expressions were presented. With the complex forms generated with Fourier series, or truncations thereof, it should be possible to generate very complex curves when they are combined from the parametric components of a plane curve.

Because there is considerable variety in this approach, what basic design goals should we consider to keep the illustration of this approach reasonably compact? Let us, for the purposes here, constrain ourselves to either symmetric or antisymmetric curves. Using the parametric form, symmetric curves require that $f(t)$ be antisymmetric and $g(t)$ be symmetric; antisymmetric curves require that $f(t)$ again be antisymmetric but that $g(t)$ be antisymmetric also. Let us also use the truncated Fourier synthesis formula with only a few terms.

An example of a symmetric curve generated with this approach is shown in Figure 13.2.13. For this example, 101 points were synthesized from the fundamental-period harmonics and the 5 higher harmonics, each having a unit amplitude coefficient. In order to keep the curve from looping back across the y axis, it is necessary to add a linear term to the values of x generated in this way so that they increase away from the y axis in both directions. The linear term is negative for the left half of the x values and positive for the right half. The antisymmetric version of the same curve is produced when the cosine series for y is changed to a sine series; it is plotted in Figure 13.2.14.

In the above syntheses of parametric curves, the Fourier series coefficients were all unity. What might happen if we allow them to become random, following the approach in Section 13.2.4? The result should certainly be less regular. Let us use

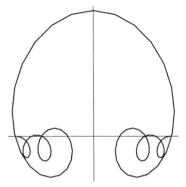

FIGURE 13.2.13. Parametric curve generated from cosine harmonics.

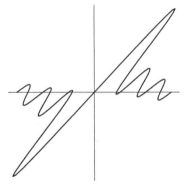

FIGURE 13.2.14. Parametric curve generated from sine harmonics.

random coefficients in the range of $-1/2$ to $+1/2$ for the Fourier series and otherwise synthesize the curves as previously done with the constant coefficients. Figure 13.2.15 shows a symmetric curve generated in this way. Figure 13.2.16 shows another curve generated in the same manner except the y values are made antisymmetric by using a sine series.

If we abandon the requirement that the result be symmetric or antisymmetric and instead use Fourier series of both sine and cosine terms for each of the components x and y, the results may still be useful. The five curves in Figure 13.2.17 were generated by using the first ten cosine and sine terms of the Fourier series for both the x and y components. The amplitude coefficients A_n and B_n were randomized to lie between $-1/2$ and $+1/2$. These examples are perhaps reminiscent of human doodling.

Because of the randomness of this approach, curves generated in this way are little more than interesting curiosities. The likelihood is that we would need to generate numerous realizations of such curves before getting a useful result for a particular application. However, it should be recognized that there is some middle ground between uniform Fourier coefficients and random ones. This is the topic of the next section.

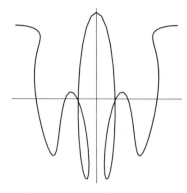

FIGURE 13.2.15. Parametric curve generated from random-amplitude cosine harmonics.

FIGURE 13.2.16. Parametric curve generated from random-amplitude sine harmonics.

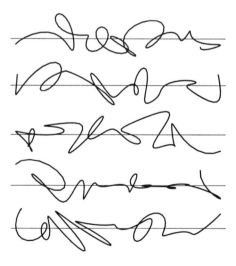

FIGURE 13.2.17. Random parametric curves generated from harmonics.

13.3. FILTERING TO MODIFY CURVES

13.3.1. Concept of a Filter

Previous parts of this chapter have shown that curves can be composed of a series of harmonic waves. The amplitudes of the spectral components are together called the amplitude spectrum of the curve, and the phases of these components are the phase spectrum of the curve. We will define a *filter*, in this context, to be a spectral multiplier that changes either the amplitudes or phases in some manner dependent upon frequency. By changing the spectrum, a consequent change in the appearance of the curve must occur. Let the filter be given by $H(f)$ in general. Then, if the spectrum of the curve is $G(f)$, the spectrum of the filtered curve is $H(f)\,G(f)$; and this can be inverted back to the space domain to produce the filtered curve.

Some types of filters are commonly used and have well-known names. For instance, an *all-pass filter* is one which does not change the amplitude spectrum. On the other hand, a *phaseless filter* is one which affects the amplitude spectrum but does not change the phase spectrum. We have already presented the *Hilbert transform filter* in Section 13.2.1, which changed the phase by $\pi/2$ when we switched the cosine coefficients to sine coefficients (or vice-versa). A *low-pass filter* attenuates the high-frequency components and a *high-pass filter* attenuates the low-frequency components. There is also a *bandpass filter* which passes a middle portion of the spectrum while attenuating the higher and lower frequencies.

This filtering operation bears little relation to the transformation operations discussed in Chapter 10. The operators discussed there applied to all curves in a geometric sense, involving stretching, contracting, rotation, shearing, and similar operations without regard to frequency content. Filtering does not ordinarily have a clear geometrical interpretation.

Filter theory has deep roots in the fields of electronics and signal processing. Filters are used to diminish noise, to shape signals to desired forms, and to enhance certain portions of the spectrum. Many of these objectives can be carried over to curve design. The following sections will present some of the possible applications.

13.3.2. Low-Pass Filtering of Curves

The generic low-pass filter is given by

$$H(f) = \frac{1}{(1 + f/f_c)^m}$$

where f is the variable frequency, f_c is some fixed corner frequency, and m is a positive integer. The integer m is often referred to as the number of *poles* in the filter. The filter relation at high and low frequencies is rather simple. This follows from taking the logarithm of both sides of the filter equation above to get

$$\log[H(f)] = -m\log[1 + (f/f_c)]$$

This evaluates to the following asymptotes for small and large f:

$$\log[H(f)] = 0 \qquad f \ll f_c$$
$$\log[H(f)] = -m\log(f) \qquad f \gg f_c$$

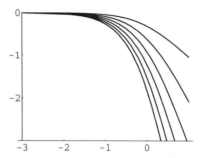

FIGURE 13.3.1. Low-pass filter for $m = 1, 2, 3, 4, 5, 6$.

At low frequencies, the logarithm of the spectrum is zero, meaning the spectral amplitude is unity. At high frequencies, the decay of the logarithm of the spectrum is linear with the logarithm of the frequency. Consider this filter's effect on a white spectrum whose maximum frequency is 10. ("White" in spectral terminology means "of uniform amplitude.") If we set $f_c = 1$, one-tenth the maximum frequency in the spectrum, the spectral amplitude of this filter for values of $1 \leq m \leq 6$ appears as shown in Figure 13.3.1. The amplitude and frequency are plotted with a logarithmic (base 10) scale, as is customary in engineering, to reflect the linear high-frequency relation of the logarithmic values.

Now frequencies of the Fourier terms which we have been using to synthesize curves. The first is the fundamental frequency f_1, equal to $1/L$ where L is the length of the curve segment to be synthesized, as measured by the domain of the independent variable. The second, and so forth, frequencies are multiples (harmonics) of this fundamental frequency thus:

$$f_n = nf_1 \quad n = 2, 3, 4, \ldots$$

The amplitude of the nth harmonic is then

$$H(f_n) = \frac{1}{\left[1 + nf_1/f_c\right]^m}$$

and that of the $n + 1$ harmonic is

$$H(f_{n+1}) = \frac{1}{\left[1 + (n+1)f_1/f_c\right]^m}$$

Provided f_1 is beyond the corner frequency f_c, the ratio of the amplitudes of the n and $n + 1$ harmonic can be reduced to approximately

$$H(f_n)/H(f_{n+1}) \approx [(n+1)/n]^m$$

In order to illustrate the effects of filtering the spectrum, we first generate a 101-point symmetric wavelet of length two units using the first eight cosine terms of the Fourier series for $-1 < x < 1$ with the coefficients of the terms all set equal to unity. This wavelet is shown in Figure 13.3.2. Now, we set the corner frequency of the

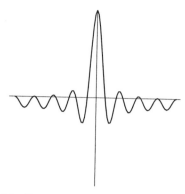

FIGURE 13.3.2. Symmetric wavelet from truncated white spectrum.

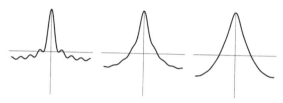

FIGURE 13.3.3. Low-pass versions of symmetric wavelet ($m = 1, 2, 3; f_c = f_1$).

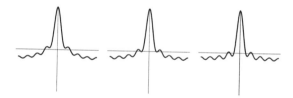

FIGURE 13.3.4. Low-pass versions of symmetric wavelet ($f_c = 1/2, 1, 2f_1; m = 1$).

low-pass filter equal to the fundamental frequency and compute the coefficients of the Fourier cosine series based on the amplitude spectrum of this filter. The wavelet was regenerated with these coefficients using $m = 1, 2$, and 3; the results are shown in Figure 13.3.3. Note that the low-pass filter removes more and more of the high-frequency content as m is increased. The other parameter that can be varied in the filter is the corner frequency. The group of wavelets in Figure 13.3.4 shows the effect of using $f_c = 1/2, 1$, and 2 times the fundamental frequency, and m is held constant at 1. As f_c increases, more of the high frequencies are preserved.

In the cases presented so far in this section, we always started with a white (truncated) spectrum. This is not required, and any spectrum can be multiplied with the filter response. We now create an example of low-pass filtering using random spectral amplitudes. The first eight terms of the sine and cosine series are used, and their amplitudes are randomized by taking numbers from a range of $-1/2$ to $+1/2$ assuming a uniform probability density. A curve is then generated from these components, and a particular realization is shown in Figure 13.3.5.

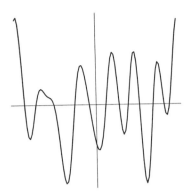

FIGURE 13.3.5. Curve from random sine and cosine components.

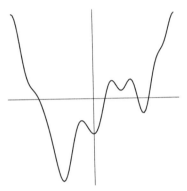

FIGURE 13.3.6. Low-pass filtered curve from random sine and cosine components.

We now apply a low-pass filter with $f_c = f_1$ and $m = 2$ to the coefficients of the sine and cosine series and regenerate the curve. The result, shown in Figure 13.3.6, shows predominantly the low-frequency information of the original curve.

It should be clear that we can compute the spectrum of any arbitrary curve, apply the filtering to the spectral coefficients, and transform back to the space domain to produce a "filtered" curve. The degree of filtering is controlled by the filter parameters: the corner frequency f_c and the exponent m of the rolloff beyond this corner.

13.3.3.   High-Pass Filtering of Curves

We saw that low-pass filtering removed the high-frequency content of a typical wavelet. However, the high frequencies often provide interesting shapes to curves. Therefore, it is desirable in many cases to do high-pass filtering of the spectrum.

The generic high-pass filter is given by

$$H(f) = \frac{1}{\left[1 + f_c/f\right]^m}$$

This is identical to the expression for the low-pass filter except f_c/f is the reciprocal of that used for the low-pass filter. The filter relation at high and low frequencies

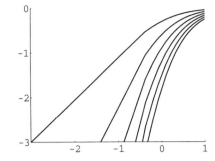

FIGURE 13.3.7. High-pass filter for $m = 1, 2, 3, 4, 5, 6$.

FIGURE 13.3.8. High-pass versions of symmetric wavelet ($m = 1, 2, 3; f_c = 4f_1$).

is rather simple. This follows from taking the logarithm of both sides of the filter equation above to get

$$\log[H(f)] = -m \log[1 + (f_c/f)]$$

This evaluates to the following asymptotes for small and large f:

$$\log[H(f)] = m \log(f) \quad f \ll f_c$$
$$\log[H(f)] = 0 \quad\quad\;\; f \gg f_c$$

At low frequencies, the increase of the logarithm of the spectrum is linear with the logarithm of frequency. At high frequencies, the spectral amplitude is constant. As for the low-pass filter, consider this filter's effect on a white spectrum whose maximum frequency is 10. If we set $f_c = 1$, one-tenth the maximum frequency in the spectrum, the spectral amplitude of this filter for values of $1 \le m \le 6$ is shown in Figure 13.3.7 in log-log coordinates.

Consider the basic "white" wavelet used in the previous section. Its length is two units, and the first eight terms of the cosine series were used to generate it. If we set $f_c = 4f_1$ in the high-pass filter expression and let m range from 1 to 3, the resulting filtered wavelets are shown in Figure 13.3.8. The increase in high-frequency is not great as m increases, but it can definitely be seen. If we let $m = 4$ and allow f_c to be 2, 4, and 8 times f_1, the wavelets appear as shown in Figure 13.3.9. Again, the differences are apparent but small.

The above has shown how a basic wavelet responds to high-pass filtering. Now we look at how a random curve responds to the same filtering. The random curve is generated with the first eight terms of both the sine and cosine series, each having

FIGURE 13.3.9. High-pass versions of symmetric wavelet ($m = 4$; $f_c = 2, 4, 8f_1$).

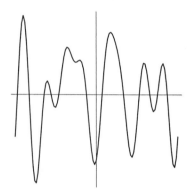

FIGURE 13.3.10. Curve from random sine and cosine components.

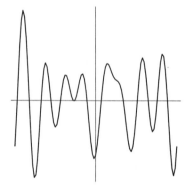

FIGURE 13.3.11. High-pass filtered curve from random sine and cosine components.

a random amplitude coefficient. The random amplitudes are uniformly distributed between $-1/2$ and $+1/2$. A curve generated in this way is shown in Figure 13.3.10. Now the high-pass filter with $f_c = 8f_1$ and $m = 2$ is applied to both the sine and cosine spectral coefficients, and the curve is regenerated from the "filtered" series. The new curve is shown in Figure 13.3.11; compared with the one in Figure 13.3.10, this curve emphasizes the high frequencies.

13.3.4.   Applications to Parametric Curves
The filtering process can be extended to affect one or both parts of the parametric curve given by $x = f(t)$ and $y = g(t)$. Let us consider a curve such as the one generated at

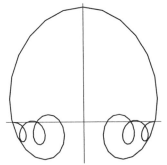

FIGURE 13.3.12. Parametric curve from harmonics.

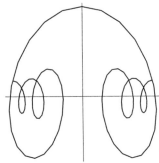

FIGURE 13.3.13. High-pass filtering of y component of parametric curve from harmonics.

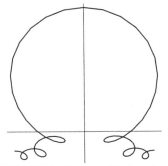

FIGURE 13.3.14. Low-pass filtering of y component of parametric curve from harmonics.

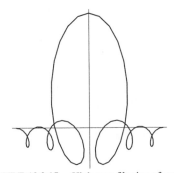

FIGURE 13.3.15. High-pass filtering of x component of parametric curve from harmonics.

the beginning of Section 13.2.5; it is repeated in Figure 13.3.12. If the low-frequency content of the y component of this curve is deemed to be too great, then a high-pass filter may help to shape the curve towards something more desirable. To this end, such a filter with $f_c = 8f_1$ and $m = 2$ was applied to the Fourier coefficients for the y component and the x component was unchanged; the regenerated curve is shown in Figure 13.3.13. Note that the smaller features, as seen in the loops, are enhanced relative to the overall extent of the curve.

Similarly, if we desired to suppress the high-frequency content of the y component, a low-pass filter is needed. To this end, such a filter with $f_c = f_1$ and $m = 2$ was applied to the y component; the result is shown in Figure 13.3.14. The character of the curve has changed significantly with this operation.

Next we will apply the filters to the x component. Again recall the basic curve shown in Figure 13.3.12 prior to filtering. If the high-pass filter is applied to the x component rather than the y component, using $f_c = 8f_1$ and $m = 2$, the result is as shown in Figure 13.3.15. Figure 13.3.16 shows the effect of a low-pass filter on the x component of the original curve. Filter parameters of $f_c = f_1$ and $m = 2$ were used. Note that the loops have disappeared due to the decrease of high-frequency content in the x component.

With just this one example, we can see that the application of basic low-pass or high-pass filters to the spectral amplitude will change the parametric curve significantly

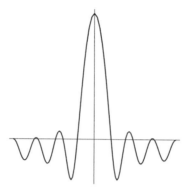

FIGURE 13.3.16. Low-pass filtering of x component of parametric curve from harmonics.

and in predictable ways. One or both of the components can be filtered to achieve a desired effect. This method can be applied to any parametric curve, whether generated from a known function, generated randomly, or perhaps constructed with splines. The application of filtering considerably broadens the base of curve shapes which can be constructed via functional means. It makes possible the reshaping of curves which have no mathematical function as their source because the Fourier series of any two-dimensional curve can always be calculated, filtered, and inverted back to a new curve.

13.4. SUMMARY

Finite, sampled curves can always be represented by a sum of sine and cosine terms; this is the essence of the Fourier series representation. Infinite curves can also be represented, but only within a finite portion of their length. Curves not ordinarily considered to be made up of harmonic components, such as sawtooth functions, can be remarkably well approximated by the Fourier series too. The Fourier series constitutes a special set of basis functions and may be compact, and thus efficient, for many types of curves.

A useful approach to the modification or reshaping of a curve is through the Fourier spectrum of that curve. Once the spectrum is computed (Fourier analysis), various filters can be applied to the spectrum to enhance or suppress certain frequencies; the filtered spectrum is then inverted back to a new curve (Fourier synthesis). One particularly simple filter is a phase change of $\pi/2$, which switches a symmetric curve to an antisymmetric one or vice-versa. Other useful filters are of the low-pass and high-pass kinds, with respect to the frequencies of the spectrum.

Chapter 14

COMPLEX CURVES FROM WAVELETS

14.1. INTRODUCTION TO WAVELETS

Chapter 13 showed how we can use the Fourier series, composed of sine and cosine terms, to synthesize complex curves in two dimensions. It was also shown that the inverse, decomposing an arbitrary curve into a Fourier series, is possible for arbitrarily complex curves. Fourier theory has held a long, useful, and dominant role in the analysis and synthesis of signals for science and engineering and was a logical choice for our study of complex curves. However, there are other approaches to the decomposition of signals that have been used with recent success. Chief among these is a decomposition method based on wavelets.

Recall that sine or cosine waves, the basic functions for Fourier theory, have infinite duration and are periodic. Thus, they appear identical no matter where they are viewed in time (or space). On the contrary, a wavelet has finite duration and is localized to a certain point in time for time-dependent signals or to a certain point in space for space-dependent curves. Like sine or cosine waves, wavelets can take on a scale relative to the axis on which they are drawn. The sine wave

$$\sin(x/b)$$

has the scaling factor b by which it can be stretched or compressed along the x axis. (Note that the factor b is in the denominator rather than in the numerator; either is usable, but the present convention conforms to wavelet theory.) Without yet giving a particular form to a wavelet, we simply denote it by $w(x)$; and so the general wavelet

$$w(x/b)$$

will be stretched or compressed according to the factor b. This factor is usually called the *dilation factor* in wavelet theory. We can readily grasp that a wavelet analysis will necessarily include wavelets of varying scale (dilation) just as the Fourier analysis includes harmonic functions of varying scale (frequency).

The important difference between wavelet analysis and Fourier analysis is in the localization of the wavelet components. Recall that a sine (or cosine) wave can be shifted by a phase factor ϕ; this can be expressed with

$$\sin(x/b + \phi)$$

This does not, however, affect the fact that the sine wave is infinite in extent. A shift of exactly 2π will reproduce the original sine wave. For the wavelet, we can also introduce a shift term (often called translation) with

$$w[(x - a)/b]$$

This wavelet is identical in appearance to $w(x/b)$, but the shift of a units enables great flexibility in wavelet analysis and synthesis because the wavelet components can be

225

localized to the exact locations given by *a*. Signals or curves that have great variability in their frequency and amplitude over time are good candidates for wavelet analysis. In time-series terminology, such signals are called *nonstationary* and have always presented special problems for the Fourier approach. In terms of basis functions as discussed in Section 11.1, the wavelet functions may provide a more compact basis for certain curves than harmonic functions.

Wavelet analysis is a well-developed formal subject, but it is not as easy to present as Fourier analysis. We only give the main result of the theory here. Consider a simple single-valued curve $y = f(x)$. The fundamental result from wavelet theory is that an arbitrary function y over an infinite duration of x can be expressed as

$$y = k \int_{-\infty}^{\infty} \int_{-\infty}^{\infty} [g(a,b)w(a,b,x)/a^2]da\,db$$

where *b* and *a* are the dilation and translation factors, respectively; and *k* is a constant dependent on the exact choice of wavelet function. (The reader unfamiliar with integrals can just regard these expressions as a sort of summation.) The amplitude scale factor *g* depends on *a* and *b*. This equation states that an integration over varying *a* and *b*, with the wavelet weighted appropriately, will produce the function *y*.

For our purposes, the integral form above is of little use. We would like to generate arbitrary functions from wavelets, but using only discrete values of *a* and *b* rather than an infinite continuum. It can be shown in wavelet theory that a function over a finite extent can be represented by an infinite series of wavelets with scale and dilation factors which change in a uniform manner, much as the frequency of a discrete Fourier series is incremented uniformly from term to term in the series. The discrete wavelet representation is then given by

$$y = k \sum_{m=1}^{\infty} \sum_{n=1}^{\infty} g_{m,n}w_{m,n}(x)$$

where *m* and *n* indicate the discretization of the coefficients *a* and *b*. In practice, as for Fourier synthesis, the summation will extend only through a finite *m* and *n*.

It is common, but not required, in wavelet analysis to make the dilation factor change in steps of two from its smallest value. Thus, if the fundamental value of *b* equals two, then allowable values of *b* are $2, 4, 8 \ldots$ This is different than the Fourier analysis where each Fourier component has a frequency that is an integer multiple of the smallest, or fundamental, frequency, with all integers used from one to infinity. On the basis of this choice for *b*, the translation values *a* are then simply $a = 2, 4, 8 \ldots$ Therefore, the discrete wavelets in the above can be expressed fully as

$$w_{m,n}(x) = 2^{-m/2}w(2^{-m}x - n) \quad (m = 1, 2, 3 \ldots; n = 1, 2, 3 \ldots)$$

The above expressions have defined wavelet representations using infinite integrals or infinite series. For tractable computation, we will use truncated series of wavelets in the following sections, just like we used truncated Fourier series in Chapter 13 to generate functions. It will be revealed that even a small number of wavelets, suitably dilated, translated, and scaled, will create interesting two-dimensional curves.

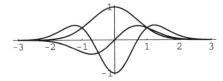

FIGURE 14.2.1. Gaussian curve and its first two derivatives.

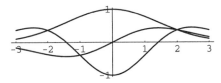

FIGURE 14.2.2. Gaussian family of curves dilated by two.

14.2. A FAMILY OF WAVELET FUNCTIONS

The types of wavelets that satisfy the criteria required by wavelet theory are often unusual and unfamiliar to those interested in curve design issues. Consequently, we will focus on one familiar family of wavelets which are, in spite of their simplicity, highly useful. In Chapter 4, the Gaussian curve was introduced, along with its derivatives. The equations of the Gaussian and its first two derivatives were

$$\exp(-x^2)$$
$$-2x\exp(-x^2)$$
$$(-2+4x^2)\exp(-x^2)$$

These curves are plotted again in Figure 14.2.1. Note that the Gaussian curve itself is entirely positive while the two derivatives have both positive and negative portions. Also note that the Gaussian curve and the second derivative are symmetric while the first derivative is antisymmetric. We will make use of these symmetry properties in the sections that follow. Note that this family of curves has the most desirable feature of wavelets: they are well localized about the midpoint, having insignificant amplitude away from it. This feature is a result of the exponential decay of these curves.

In order to use these curves in what follows, the variable x will be dilated by the parameter b and the wavelet will be translated by the amount a so that $(x - a)/b$ replaces x in the wavelet to be summed into the curve. The effect of changing the translation term is clear: the wavelet is moved to the left or right, depending on whether a is negative or positive, respectively. The effect of increasing the dilation factor may not be so clear. In the above plot, the dilation factor is simply unity; if it is increased to a value of two, the curves appear as in Figure 14.2.2.

Therefore, we can see that increasing the dilation factor stretches the curves out. In order to use these curves as wavelets, it will be advantageous to weight them such that they are entirely localized to the interval $[-d, d]$ and equal to zero at both endpoints. A weighting function that can achieve this goal is $\cos[\pi x/(4d)]$; values outside the interval $[-d, d]$ are set to zero. This function would appear to have a bell shape

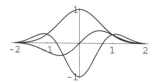

FIGURE 14.2.3. Gaussian family of curves shaped by cosine bell curve.

FIGURE 14.3.1. Gaussian wavelets joined together.

FIGURE 14.3.2. Sine wave with added linear term.

between $-d$ and $+d$. When multiplied by this bell-shaped curve, the three wavelets appear as in Figure 14.2.3. There are only slight differences between these truncated, weighted wavelets and the original functions in Figure 14.2.2. These completely localized wavelets will be used in the following sections while still being referred to as the Gaussian wavelet or its first or second derivative.

14.3. DETERMINISTIC CURVES GENERATED FROM WAVELETS

Recall that two-dimensional curves can be generated with the parametric representation $x = f(t)$ and $y = g(t)$. In our first attempt at a wavelet-generated curve, let us represent y with wavelets while using the more familiar sine function for x. Figure 14.3.1 shows y created from nine Gaussian wavelets with parameters $a = -4, -3, -2, -1, 0, 1, 2, 3, 4$ and all $b = 0.3$. (The axes have been omitted; the y axis bisects the curve while the t axis is slightly below the entire curve.) This resembles a sine wave quite closely. Now let

$$x = \sin(2\pi t)/10 + t/5$$

This curve is shown in Figure 14.3.2. This sine wave, plus a linear term, is synchronized with the wavelet series in Figure 14.3.1 such that each cycle of the sine wave coincides with each wavelet in time. Recall from Chapter 5 that the coefficient of the linear term controls the overlapping of the loops: a lower coefficient gives more overlap while a higher value will tend to remove overlap. Combining the x and y parametric parts gives the curve shown in Figure 14.3.3; not surprisingly, it resembles a cycloid. Consequently, at this point, we have achieved little more than what could be done with only harmonic functions.

If the previous curve is repeated, but now changing b from 0.3 to 0.2, the curve shown in Figure 14.3.4 is the result. The "flattened" parts of the curve at the bottom are due to the fact that the Gaussian wavelets for the y part are compressed, leaving significant portions of near-zero values between the individual wavelets. To achieve an opposite affect, the curve is regenerated, but with $b = 0.6$ now, as shown in Figure 14.3.5. This result is somewhat smoother than the original in Figure 14.3.3 with

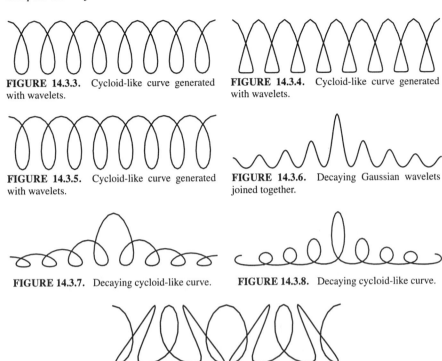

FIGURE 14.3.3. Cycloid-like curve generated with wavelets.

FIGURE 14.3.4. Cycloid-like curve generated with wavelets.

FIGURE 14.3.5. Cycloid-like curve generated with wavelets.

FIGURE 14.3.6. Decaying Gaussian wavelets joined together.

FIGURE 14.3.7. Decaying cycloid-like curve.

FIGURE 14.3.8. Decaying cycloid-like curve.

FIGURE 14.3.9. Decaying cycloid-like curve.

$b = 0.3$. The point to recognize here is that, by merely dilating the wavelet, some control is exercised over the shape of the curve. This control is difficult to achieve with harmonic functions alone, even with the summation of many harmonic terms.

Using the same values of the translation terms a and $b = 0.3$ again, let us now scale the amplitudes of the individual wavelets according to the factor $1/(|a| + 1)$ which will cause them to decay on each side of the y axis relative to the central one. Figure 14.3.6 shows the plot of y itself. When this y function is combined parametrically with the same x function as plotted in Figure 14.3.2, the curve shown in Figure 14.3.7 is created. It is interesting that a different, but equally pleasing, curve can be created simply by changing the sign of the sine term in the x function as seen in Figure 14.3.8.

Many variations of the above could be created by using different decay factors, by changing the dilation factor, or by introducing other wavelets such as the second derivative of the Gaussian curve. Another interesting variation can be created by making the x component asynchronous with the y component. The curve in Figure 14.3.9 was generated using the original y component of this section but changing the coefficient of the argument of the sine wave in the x component from 2.0 to 2.5.

Thus far, a constant dilation factor has been used in all the wavelets of the y component. We now add more wavelets with a different dilation factor. In changing the dilation factor, the spacing of the wavelets is accordingly increased or decreased to accommodate the new size of the wavelet. (This is not a requirement.) Take the original y component of this section which used $b = 0.3$ and had nine wavelets; now

FIGURE 14.3.10. Curve with mixed, decaying wavelets.

FIGURE 14.3.11. Parametric curve with mixed, decaying wavelets.

FIGURE 14.3.12. Parametric curve with mixed, decaying wavelets.

FIGURE 14.3.13. Parametric curve with mixed, decaying wavelets.

FIGURE 14.3.14. Parametric curve with mixed, decaying wavelets.

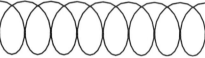

FIGURE 14.3.15. Parametric curve with Gaussian wavelets for x and y.

add a new set of wavelets with $b = 0.6$. Let the translation terms of this new set be $a = -4, -2, 0, 2, 4$ to give five wavelets. These terms increase the spacing in proportion to the increase in b. Let the amplitude scaling of the individual wavelets be given by $1/(|a| + 1)$ again. The y component appears as shown in Figure 14.3.10. Combining this with $x = \sin(2\pi t)/10 + t/5$ again gives the parametric curve shown in Figure 14.3.11. A trivial change in scale factor to $1/(|a| + 3)$ produces the curve shown in Figure 14.3.12. If the scale factor of the second set of wavelets is negative rather than positive while keeping everything else the same, the curve shown in Figure 14.3.13 is the result.

We are not limited to symmetric curves. By switching from the Gaussian wavelet to its first derivative for the y component and keeping everything else the same from the previous symmetric curve of Figure 14.3.13, the antisymmetric curve of Figure 14.3.14 is generated.

Clearly, the variations are unlimited with just this simple use of wavelets in the y component as done in this section. This is due to the ability to localize their effect via translations and to individually dilate and scale them.

So far, we have used wavelets in the y component but harmonics in the x component. In the rest of this section, we will also create the x component from wavelets; but it should be recognized that the required x wavelet in this case must be antisymmetric to maintain the symmetry of the resultant parametric curve. Thus, our choice of wavelet for the x component will be the first derivative of the Gaussian function.

The first curve to be generated uses $b = 0.4$ and $a = -4, -3, -2, -1, 0, 1, 2, 3, 4$ for nine wavelets in both the x and y components. The y wavelet is the Gaussian curve and the x wavelet is its first derivative. We form the x component first with the wavelets, and then the term $t/2$ is added for $-5 < t < 5$. The result is shown in Figure 14.3.15. Note that it appears similar to a cycloid. By varying the value of b

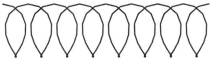

FIGURE 14.3.16. Parametric curve with Gaussian wavelets for x and y.

FIGURE 14.3.17. Parametric curve with mixed Gaussian wavelets for x and y.

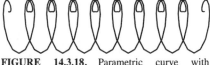

FIGURE 14.3.18. Parametric curve with wavelets placed end-to-end.

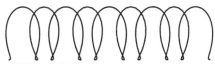

FIGURE 14.3.19. Parametric curve with mixed wavelets placed end-to-end.

in either or both of the components, the cycloid-like curve can be easily changed in appearance. If the wavelets are compressed using $b = 0.25$ instead, the curve shown in Figure 14.3.16 is the result. The next example starts with $b = 0.4$ for the same values of a but also adds dilated wavelets with $b = 0.8$ at $a = -4, -2, 0, 2, 4$ for both the x and y components. The result is shown in Figure 14.3.17.

A flaw with the approach used so far is that, if the wavelets are compressed too much by making b small, there are significant ranges of x or y over which the influence of the wavelets is negligible. This leaves large, uninteresting, flat or linear portions in the parametric curve at regular intervals. Thus far, a deliberate attempt was made to avoid such situations. If we no longer try to follow the wavelet theory in any manner, a modification of the synthesis approach is possible to guarantee that we produce pleasing curves in every case: let the wavelets simply join end-to-end. In effect, the translation term now depends on the dilation factor.

The first curve to be constructed in this manner uses nine wavelets for both the x and y components. The x wavelet is the first derivative of the Gaussian, and the y wavelet is the second derivative of the Gaussian. The value $b = 0.3$ is used for both wavelets, and $t/50$ was added to the x component. Figure 14.3.18 shows the result.

The real value of the end-to-end technique is that now we can intermix wavelets of differing b. Let the y component wavelet be the Gaussian curve and the x component wavelet be its first derivative. Using $b = 0.2$ and $b = 0.4$ for alternate wavelets in the series and adding $t/75$ to the x component produces the curve shown in Figure 14.3.19. The effect of the alternating wavelet sizes can be clearly seen here.

It is interesting to control the dilation and scale of the wavelets even more. In the next example, nine wavelets are again used, but each is compressed and scaled relative to the center one. With the wavelets indexed by $i = -4, -3, -2, -1, 0, 1, 2, 3, 4$, set the dilation factor $b = (5 - |i|)/10$ and the scale factor c to the same value. Again putting the wavelets end-to-end and adding t/s to the x component gives the curves in Figure 14.3.20 for $s = 25, 75$, and 200.

As a last example this section, let us repeat the above curves but use the second derivative of the Gaussian curve in the y component instead of the Gaussian curve itself and change the s values to $s = 100, 200$, and 400. The curves then appear as shown in Figure 14.3.21.

It would be very difficult to achieve curves similar to those in this section using only

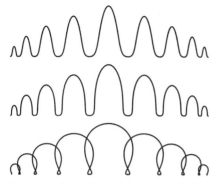

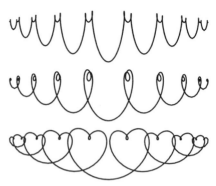

FIGURE 14.3.20. Parametric curves with dilating wavelets placed end-to-end.

FIGURE 14.3.21. Parametric curves with dilating wavelets placed end-to-end.

FIGURE 14.4.1. Random curve from wavelets.

FIGURE 14.4.2. Random curve from wavelets.

harmonics. The wavelet approach gives considerable flexibility to the generation of curves, and these curves can be controlled in a very localized manner as demonstrated by the examples given here.

14.4.   RANDOM CURVES GENERATED FROM WAVELETS

In Chapter 13 curves were generated with random amplitudes and phases for harmonic components. Here, we will try a similar approach using wavelets. The wavelets will be put end-to-end while we take their dilation factor from a uniform density in the range [0.25, 1.25] and their scale factor from a uniform density in the range [−0.5, 0.5]. These random factors are chosen independently for the x and y components; but the wavelet length in both cases is taken to be that of the longer wavelet with the shorter one augmented by zero values. A term equal to $t/400$ is finally added to the x component. One realization of such a scheme is shown in Figure 14.4.1 when 10 wavelets are used. This curve is clearly random, much like an aimless doodle; and generation of such curves seemingly serves little purpose.

In order to organize the curve somewhat, let us coordinate the dilation factors to be the same for the x and y wavelets and still keep the scale factors independently random. Figure 14.4.2 shows a realization generated with this scheme. If we further equate the scale factors between the x and y wavelets, the result becomes even more ordered as seen in Figure 14.4.3 for one realization of this scheme.

All of the above realizations used a random scale factor in the range [−0.5, 0.5]. If this range is restricted to negative only [−1, 0] while using a scheme otherwise identical to that of Figure 14.4.3, the resulting realization, seen in Figure 14.4.4, appears to have its parts homogeneous in shape but only differing in size.

FIGURE 14.4.3. Random curve from wavelets.   **FIGURE 14.4.4.** Random curve from wavelets.

The random approach to curve generation with wavelets could be varied by many other subtleties. However, the utility of the results are not considered significant enough to merit detailed pursuit. However, you may be encouraged to experiment simply by seeing the limited results here.

14.5. SUMMARY

The theory of wavelets provides us with an alternative to polynomials and harmonics for basis functions in curve generation. The wavelets are parameterized by 1) a dilation factor, which increases or decreases its length, 2) a scale factor, which controls its amplitude, and 3) a translation term, which determines where in space it will be placed. Due to the fact that finite wavelets can be localized to particular portions of the curve and that they can be independently scaled and dilated, wavelets offer possibilities unattainable with polynomials or harmonics, both of which are infinite in extent. Although many useful wavelets have been defined and employed in wavelet theory, a particularly easy one to use is the Gaussian curve, or its derivatives, weighted such that it is zero outside a given interval.

Numerous examples of curves generated with wavelets showed just a small view of the range of shapes that can be created. Symmetric or antisymmetric curves can be generated by switching between symmetric and antisymmetric wavelets as required. We explored some of the obvious variations such as changing the dilation factors in a given manner, uniformly increasing or decreasing the scale factors of successive wavelets, and changing the signs of the wavelets. Interesting results are attained when both the x and y components of parametric curves are composed of wavelets.

Chapter 15

SPACE CURVES

15.1. INTRODUCTION TO SPACE CURVES

When the points of a curve do not occupy a single plane, then its graph must occupy space. Such curves are called *space curves*, also *three-dimensional curves*. The development of plane curves can be extended to space curves easily. The behavior of polynomial and transcendental components of the curve can still be predicted and analyzed, and there is effectively nothing conceptually new to present in regard to space curves. Properties such as periodicity, continuity, and symmetry are easily applied to these curves along lines of definitions that have already been made.

In order to treat space curves, we must introduce another coordinate to define a three-dimensional space. Plane curves were defined in the Cartesian coordinates x and y (or the polar coordinates r and θ). For three-dimensional, we extend the Cartesian coordinates with z; we will show these as a box projected to the viewing plane of the page. There are many ways to handle the projection of three-dimensional space to a two-dimensional viewing plane (none of which are entirely satisfactory). The method used here is called the "perspective" projection. Imagine a viewer at some distance from a plane that is perpendicular to the line of sight. Place the plane so that it passes through the origin (0, 0, 0) of the three-dimensional object. The points of the object are then projected along the lines which connect them to the viewer and plotted on the two-dimensional projection plane where these lines intersect it. Thus, the viewer perceives, artificially, the true perspective of the object as if it were in space. Figure 15.1.1 illustrates the three-dimensional Cartesian system and this perspective projection of a unit cube, as seen from the point $(2, -5, 2)$.

Alternative coordinate systems for three-dimensional are the cylindrical and spherical coordinates. *Cylindrical coordinates* are the triplet (r, θ, z) where r is the horizontal distance from the z axis, z is the distance above or below the x-y plane, and θ is the azimuth measured identically to θ for polar coordinates in the plane (Section 1.1). Cylindrical coordinates are, therefore, just an extension of polar coordinates with the additional z variable. These coordinates are convenient for figures that are symmetric about the z axis. *Spherical coordinates* are the triplet (r, θ, ϕ) where r is the distance from the origin, θ is defined as for cylindrical coordinates, and ϕ is the angle, as measured in a vertical plane, from the positive z axis. Spherical coordinates are convenient when the figure has r defined in terms of the two angles θ and ϕ.

Space curves are normally expressed parametrically by the form

$$x = f(t)$$
$$y = g(t)$$
$$z = h(t)$$

where t is the independent variable. However, in many cases, the relationships between the three dependent variables may be found such that the space curve can be

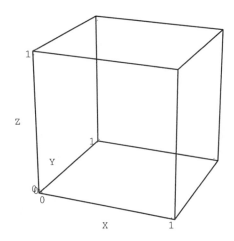

FIGURE 15.1.1. Three-dimensional Cartesian system.

expressed by just two equations; for instance

$$y = g'(x)$$
$$z = h'(x)$$

Even though such relations often give more insight into the shape of the curve prior to actually plotting it, the full parametric form is the most direct way to evaluate and plot the space curve.

As for parametric curves in the plane, the continuity of the space curve is controlled by the continuity of each of its three components. If all three components are continuous within the defined range of the independent variable t, then the space curve is continuous; conversely, discontinuity of any one of the three components is sufficient to make the space curve discontinuous.

Space curves, like plane curves, may be open or closed. If the range of t is $[t_{min}, t_{max}]$, a closed space curve must have $x(t_{min}) = x(t_{max}), y(t_{min}) = y(t_{max})$, and $z(t_{min}) = z(t_{max})$. Just as for plane curves, there may be one or more points of self-intersection or of osculation along the space curve.

Symmetry of space curves follows from the symmetry of plane curves, with an added dimension. Symmetry is now defined relative to planes rather than lines. A curve symmetric about the x-y plane has mutually reflective points everywhere given by (x, y, z) and $(x, y,$-$z)$. Similarly, symmetry about the y-z or x-z planes is defined by the existence of (x, y, z) and $(-x, y, z)$ everywhere along the curve or by the existence of (x, y, z) and $(x,$-$y, z)$, respectively. Antisymmetry about the x-y plane is defined by mutually reflective points (x, y, z) and $(-x, y, $-$z)$ or by (x, y, z) and $(x, $-$y,$-$z)$. Similarly, antisymmetry about the x-z or y-z planes can be defined. A *point symmetric* space curve has points that are mutually reflective through the origin at $(0, 0, 0)$; thus, for any point (x, y, z), a corresponding point $(-x, -y, -z)$ exists.

In Chapter 10 we showed that matrix transformations of plane curves were a means of varying or modifying the shapes of known curves. Such transformations can be extended to three-dimensional curves. Following the form of the equations used for two-dimensional in Section 10.1, we write the general transformation of three-dimensional Cartesian coordinates in space as

$$\begin{bmatrix} x' \\ y' \\ z' \end{bmatrix} = \begin{bmatrix} a & b & c \\ d & e & f \\ g & h & i \end{bmatrix} \begin{bmatrix} x \\ y \\ z \end{bmatrix}$$

(A similar relation can be written for other coordinates systems such as the cylindrical or the spherical.) The above matrix notation is equivalent to the equations

$$x' = ax + by + cz$$
$$y' = dx + ey + fz$$
$$z' = gx + hy + iz$$

Let the matrix be written as **T**. If the coordinate triplet (x, y, z) is denoted by p and the coordinate triplet (x', y', z') by p', then $p' = \mathbf{T}p$. The values of elements in **T** were given in Chapter 10 for many important and useful two-dimensional transformations; we should be able to extend these to three-dimensional transformations when required.

15.2. PLANE CURVES EXTENDED WITH A Z COMPONENT

One of the simplest ways to extend plane curves to space curves is to add a z variation while the graph of the curve is traced in the familiar x-y plane. Clearly, this can be done for any plane curve we can design. This section will treat only linear variations of the z component; this requires $z = ct$ where c is a constant. A very elementary example is the *circular helix* for which

$$x = a \cos(2\pi t)$$
$$y = a \sin(2\pi t)$$
$$z = ct$$

This can be more simply expressed in cylindrical coordinates as

$$r = a$$
$$\theta = 2\pi t$$
$$z = ct$$

The constant c controls the stretch of the helix in the z direction, and the constant a controls the radius of the helix. Figure 15.2.1 illustrates the helix for $a = 1$ and $c = 1$ as $0 < t < 4$. Note that the number of turns of the helix is given by the upper limit of t.

By adding polynomial or radical factors to the x and y expressions, we can make the curve increase or decrease in its extent as t increases. For example, consider

$$x = a \cos(2\pi t) \sqrt{t}$$
$$y = a \sin(2\pi t) \sqrt{t}$$
$$z = ct$$

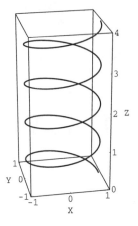

FIGURE 15.2.1. Circular helix.

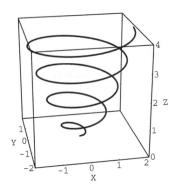

FIGURE 15.2.2. Parabolic helix.

which generates a parabolic helix as shown in Figure 15.2.2 (using the same constants as for the circular helix). Those points at the intersection of this space curve with any vertical plane including the z axis will lie on a parabola with its vertex at the origin. The projection of the trace of this space curve onto the x-y plane will give the parabolic spiral.

Either of the above figures could be stretched or compressed in x relative to y by making the constants differ for the x and y expressions; thus the x-y projected trace will be elliptical rather than simply circular.

It should be recognized that the x-y variation of the space curve determines the three-dimensional surface on which the curve lies in space. By using the constant z variation, the following correspondence exists between certain familiar x-y curves and the three-dimensional surface on which their z-extended space curve lies:

two-dimensional curve	three-dimensional surface
circle	cylinder
Archimedes' spiral	cone
Fermat's spiral	paraboloid

Any plane curve can be given the above treatment to make it a space curve. A class of fairly interesting shapes can be generated by using those plane curves that periodic on the circle as treated in Chapter 6. The circular helix is in fact the most elementary of this class. More complex periodic curves such as epitrochoids (Section 6.4.2) lead to space curves expressed as

$$x = d\{(a+b)\cos(2\pi t) - e\cos[2\pi(a+b)t/b]\}$$
$$y = d\{(a+b)\sin(2\pi t) - e\sin[2\pi(a+b)t/b]\}$$
$$z = ct$$

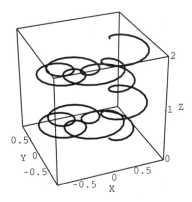

FIGURE 15.2.3. Epitrochoidal helix.

Figure 15.2.3 illustrates this three-dimensional epitrochoid with $a = 5, b = 1, c = 1, d = 0.1$, and $e = 3$ with $0 < t < 2$.

15.3. SPACE CURVES FROM A CONSTANT-LENGTH LINE SEGMENT

15.3.1. One End Fixed

This section treats space curves generated by the motion of a line segment which is of constant length a. One end of the segment will be associated with the generated trace of the space curve and will follow a prescribed motion. The other end of the segment will be fixed at $(0, 0, 0)$. With the length of the line equal to a, the parametric expressions for the space curve must satisfy

$$x^2 + y^2 + z^2 = a^2$$

Therefore, the curve lies on the surface of a sphere of radius a. An elementary family of such curves would be the loops given by

$$x = a \sin(mt) \cos(nt)$$
$$y = a \sin(mt) \sin(nt)$$
$$z = a \cos(mt)$$

Using the trigonometric identity $\sin^2(nt) + \cos^2(nt) = 1$, note that $x^2 + y^2 = a^2 \sin^2(mt)$ and so $x^2 + y^2 + z^2 = a^2$. Using $0 < t < 2\pi, m = 1$, and $n = 1$, we trace the space curve on the surface of the sphere of $a = 1$ in Figure 15.3.1; it occupies the positive y half of the spherical volume. By suitable rearrangement of the x, y, and z expressions, this space curve can be made to occupy any one of the six principal halves of the spherical volume. (The principal halves are separated by one of the three principal planes at $x = 0, y = 0$, or $z = 0$.) Figure 15.3.2 shows the same loop, but with $n = 2$. It should be realized that, in general, there are n twists in the loop.

Now let $m = 2$ and $n = 1$. The $m = 2$ factor forces the loop to osculate midway in its graph and to come back and osculate at its starting point as shown in Figure 15.3.3.

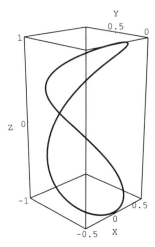

FIGURE 15.3.1. Spherical loop.

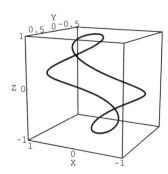

FIGURE 15.3.2. Spherical loop.

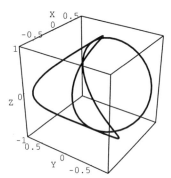

FIGURE 15.3.3. Spherical loop.

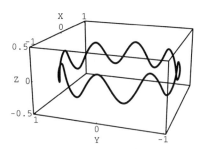

FIGURE 15.3.4. Spherical sine wave.

By increasing n, more twists can be added. If we want a deflection of z less than unity, let $z = b \cos(mt)$ where $b < a$. Using again the relation that $x^2 + y^2 + z^2 = a^2$, the components of x and y can be easily derived:

$$x = [a^2 - b^2 \cos^2(mt)]^{1/2} \cos(t)$$
$$y = [a^2 - b^2 \cos^2(mt)]^{1/2} \sin(t)$$
$$z = b \cos(mt)$$

Using $a = 1, b = 0.2$, and $m = 8$ for $0 < t < 2\pi$, the result is a sinusoidal wave which lies on the surface of a sphere and is bisected by the equator as shown in Figure 15.3.4.

These figures only give a small sample of the possible space curves that can be generated on a sphere. The key is to choose the components such that $x^2 + y^2 + z^2 = a^2$. Once the form of one of the coordinates is chosen, the remaining two are constrained to fulfill this relation. Within this constraint, however, the solution for the remaining

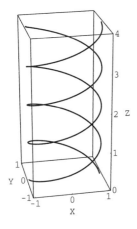

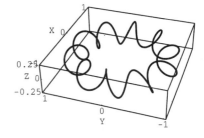

FIGURE 15.3.5. Double helix. **FIGURE 15.3.6.** Toroidal spiral.

two is not necessarily unique. In the case of Figure 15.3.4, for instance, a cycloid could be generated rather than simply the sine wave.

15.3.2. Both Ends Free

While the previous section required that one end of the line segment remains fixed, this section will treat space curves generated when this requirement is relaxed. We can generate a limitless variety of space curves under the condition that both ends move, but only a few representative ones will be treated here.

We start by noting that the circular helix in Section 15.2 belongs in this class also. Let the segment of length a rotate about the central axis; one end travels a circular path at a constant rate while the other end moves vertically up the z axis at another constant rate giving a circular helix. The cylindrical-coordinate definition of the helix seems to reflect this motion exactly: $r = a, \theta = 2\pi t, z = ct$. If the segment is lengthened to $2a$ and attached to the z axis at its midpoint, the free rotating ends each trace a circular helix. Together the figure is called the "double helix" as shown in Figure 15.3.5 for $a = 1, c = 2$ and $0 < t < 2$.

A *torus* is a solid circular ring of constant cross-section. The circular helix can be wrapped around a torus by letting the one end of the straight-line segment of length a move in a circular manner about the central axis of the torus at a rate c, and the other end of the segment moves along the central axis at a constant rate of unity. Let the central axis of the torus lie in the x-y plane. With the radius of the torus' cross-section given by a and the distance of the circular central axis from the origin given by b, the parametric equations are

$$x = [a \sin(ct) + b] \cos(t)$$
$$y = [a \sin(ct) + b] \sin(t)$$
$$z = a \cos(ct)$$

Figure 15.3.6 illustrates the *toroidal spiral* for $a = 0.2, b = 0.8$, and $c = 10$.

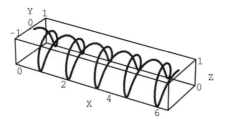

FIGURE 15.3.7. Sine wave on a cylinder.

We can make the ordinary sine wave into a space curve using the approach of this section. Consider the line segment of length a. Let one end be attached to the x axis and move along it at a rate of unity. Let the other end oscillate back and forth in the y-z direction with a sinusoidal motion like a pendulum up to an arc amplitude of b. Let α be the instantaneous angle of the line segment relative to vertical and c be the rate of oscillation; then $\alpha = (b/a)\sin(ct)$. The parametric equations describing this motion are therefore

$$x = t$$
$$y = a\sin[(b/a)\sin(ct)]$$
$$z = a\cos[(b/a)\sin(ct)]$$

Figure 15.3.7 illustrates this sine wave using $a = 1, b = \pi/2, c = 5$, and $0 < t < 2\pi$; it lies on the top half of the cylinder whose axis is coincident with the x axis and whose radius is a.

The above only represent a few of the limitless possibilities for generating space curves with a fixed line segment. Nearly any desired curve can be generated when the motion is analyzed and put into the parametric form. It is often advantageous to derive the forms based on cylindrical or spherical coordinates rather than Cartesian coordinates and then convert them to Cartesian for actual plotting.

15.4. SPACE CURVES FROM A VARIABLE-LENGTH LINE SEGMENT

When the length of the line segment is allowed to vary, the analysis of motion becomes more difficult. One of the most obvious applications of this relaxed constraint is in regard to space curves that are wrapped on a deformation of the cylinder. If the cylinder is allowed to have an ellipsoidal rather than a circular cross-section, the fixed-length line segment used to generate the helix, for instance, becomes a variable-length segment corresponding to the varying distance of the elliptical cylinder from the central axis. We can also change the helix wrapped on the torus in this manner. In such cases, the length of the generating segment varies periodically: one cycle per turn of the helix.

Another way to modify the basic circular helix is to make the variation of the length of the generator have a different rate than the rotation of the line segment about the

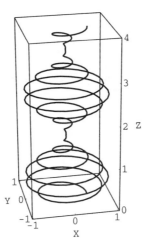

FIGURE 15.4.1. Varying circular helix.

central axis. Recall the equations of the circular helix:

$$x = a\cos(2\pi t)$$
$$y = a\sin(2\pi t)$$
$$z = ct$$

Now make the line length vary with the factor $[1 + \sin(m\pi t)]/2$, which oscillates between 0 and 1. The parametric equations are now

$$x = a\cos(2\pi t)[1 + \sin(m\pi t)]/2$$
$$y = a\sin(2\pi t)[1 + \sin(m\pi t)]/2$$
$$z = ct$$

Figure 15.4.1 illustrates this effect for $a = 1, c = 1/4, m = 1/4$, and $0 < t < 16$. We can prove that there is a node in the helix at $z = c[2n + 3/2]/m$ for $n = 0, 1, 2, \ldots$. Smaller m values will stretch out the positions of the nodes along the z axis. This one example should be sufficient to illustrate what can be achieved with variable-length line segments in generating space curves.

15.5. COMPLEX SPACE CURVES

15.5.1. A Familiar Example on a Sphere
Space curves, even though seemingly complex, can yield to analysis, especially if they are closed curves. In most cases of this type, the parametric components are composed of harmonics having fundamental periods of π or 2π. To demonstrate how we can analyze these curves, we will consider the curve that separates the two halves of familiar sports equipment: baseballs or tennis balls. You will undoubtedly have a picture of this in your mind.

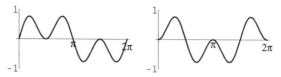

FIGURE 15.5.1. Starting curves for x and y components ($x = \sin(t)/2 + \sin(3t)/2$ and $y = \cos(t)/2 - \cos(3t)/2$).

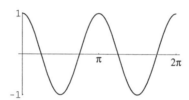

FIGURE 15.5.2. Starting curve for z component ($z = [1 - x^2 - y^2]^{1/2}$).

Orient the ball so that the ends of the flaps are facing up and down. At the top of the flap, let the x coordinate be tangential to the curve at its highest point and y be normal to it. Then the x component must start with a zero value and must be antisymmetric about this point. Letting $x = a\sin(t)$ would achieve this. However, it must have an additional higher harmonic in it to produce the correct shape of the flap. This harmonic must be $b\sin(3t)$. Now, the y component of the curve is simply $\pi/2$ out of phase with the x component, so it will be composed of the same two harmonics but changed to cosine functions. Plots of the presumed shape of the x and y components are shown in Figure 15.5.1 for $a = b = 1/2$.

We can derive the z component from the fact that the space curve lies on a sphere. Therefore, for a radius of unity, $x^2 + y^2 + z^2 = 1$ and $z = [1 - x^2 - y^2]^{1/2}$. We can show that, for the special values of $a = b = 1/2, z = \cos(2t)$; the derivation is straightforward, but it is omitted here. This z component is plotted in Figure 15.5.2 for $a = b = 1/2$.

In order to get the correct sign for z, a factor which changes between $+1$ and -1 at the zeros of z must be used. This factor has the form

$$(-1)^{\mathrm{INT}[2(t+\pi/4)/\pi]}$$

where "INT" means to take the integer part of the bracketed expression as the exponent. Using values of a and b equal to 1/2, the space curve now appears as shown in Figure 15.5.3. Note that the curve osculates at $t = 0$ and at $t = \pi$. This is due to the choice of a and b. We must point out that the choice of a relative to b is not arbitrary. To find the exact relation, consider the curve at $t = \pi/4$ where the z component equals zero. At this point, $x^2 + y^2 = 1$. By expanding the x and y components and using $t = \pi/4$, we get $a^2 + b^2 + 2ab = 1$, which reduces to $(a + b)^2 = 1$. Taking the square root of both sides, $a + b = 1$ becomes the relation for these coefficients. This must be satisfied in order that the space curve lies wholly on the sphere of radius unity. By choosing $a = 2/3$ and $b = 1/3$, the space curve plotted in Figure 15.5.4 appears to closely represent the flaps seen on the baseball or tennis ball. As a increases to unity,

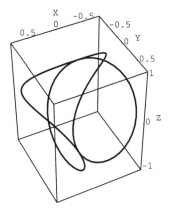

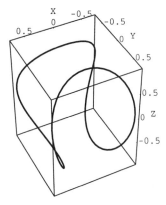

FIGURE 15.5.3. Twin spherical patches.   **FIGURE 15.5.4.** Twin spherical patches.

the curve becomes flatter and flatter until it finally becomes a circle in the *x-y* plane when $a = 1$.

For the above curve, the approach was to analyze the desired x and y components and then apply a quadratic relation, in that case the equation of a sphere, to produce the z component. In a similar way, space curves can be designed to lie on a cone, a hyperboloid, and other quadric surfaces. The following summarizes the available relations:

$$x^2 + y^2 + z^2 = 1 \qquad \text{sphere}$$
$$x^2 + y^2 - z^2 = \pm 1 \qquad \text{hyperboloid of one (+) or two (−)sheets}$$
$$x^2 + y^2 - z^2 = 0 \qquad \text{cone}$$
$$x^2 \pm y^2 - cz = 0 \qquad \text{elliptic (+) or hyperbolic (−) paraboloid}$$

By adding nonunity coefficients to the equations, we can stretch or compress these figures along any of the three principal axes.

15.5.2. Three-dimensional Lissajous Curves
One of the most interesting sets of the closed curves of Chapter 9 was the Lissajous family. The equations for generating these curves in the plane were

$$x = \sin(at + b\pi)$$
$$y = \sin(t)$$

As we saw in Chapter 9, these simple equations were able to generate a large variety of shapes. This type of curve can be extended to three dimensions by adding a third component. Let the equations for the spatial Lissajous curve be

$$x = c \sin(at + b\pi)$$
$$y = c \sin(t)$$
$$z = c \cos(t)$$

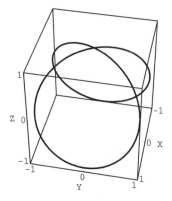

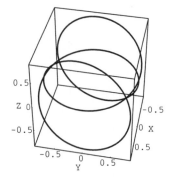

FIGURE 15.5.5. Spherical Lissajous curve.   **FIGURE 15.5.6.** Spherical Lissajous curve.

so that y and z simply define a circular motion while x introduces the complexity to the space curve. The arbitrary constant remains to be set. Note that

$$x^2 + y^2 + z^2 = c^2[1 + \sin^2(at + b\pi)]$$

Because the variable t is still in this result, the expression cannot be equal to one, or any constant, for the entire curve. Thus, the curve cannot be forced to lie on a sphere. We will arbitrarily set $x^2 + y^2 + z^2$ to unity when $t = 0$; using this value of t gives

$$c = \{1/[1 + \sin^2(b\pi)]\}^{1/2}$$

The example in Figure 15.5.5 uses $a = 1/2, b = 0$ and $0 < t < 4\pi$. The next example in Figure 15.5.6 uses $a = 1/3, b = 1/4$, and $0 < t < 6\pi$.

These examples give only a small hint of the wide variety of three-dimensional Lissajous shapes possible. Other plane curves can be given similar treatment to make complex space curves. The emphasis here has been on closed curves, but two-dimensional open curves can also lead to complex and interesting shapes when taken to three dimensions. Using the approaches exemplified here, you should be able to implement many of these possibilities.

15.6. SUMMARY

Space curves, or three-dimensional curves, involve no new concepts in comparison with two-dimensional curves. There is merely the extension that the points of the curve will all not lie within a single plane. An additional coordinate is required to specify the third dimension; and therefore, a third equation is required when the curve is specified parametrically. We can create space curves by starting with known plane curves and adding a z variation or they can be created from constant-length or variable-length lines moving in three dimensions. The equations to create many space curves which lie on familiar geometric surfaces such as spheres, cylinders, and cones are easily specified. Other space curves may require somewhat arduous analysis to specify them mathematically.

Chapter 16

SURFACES FROM CURVES

16.1. INTRODUCTION TO SURFACES

Description and analysis of surfaces is considerably more complex than for curves due to the extension from two to three dimensions. However, because only one chapter is devoted to surfaces, we will not attempt a lengthy discussion of the properties of surfaces. The primary role of this chapter is to introduce surfaces as mere extensions of curves rather than a completely new concept.

Chapter 1 introduced curves as functions in the explicit form

$$y = f(x)$$

Here y is the dependent variable, and the range of evaluation is over the x axis. Similarly, surfaces are described by the explicit form

$$z = f(x, y)$$

where the range of evaluation is now over the x-y plane. We also introduced curves in their parametric form

$$x = f(t); \; y = g(t)$$

where t is the independent parameter used for evaluation. For surfaces, the corresponding parametric form is

$$x = f(u, v); \; y = g(u, v); \; z = h(u, v)$$

where u and v are the independent parameters.

Although the explicit form or parametric form for surfaces can generate many interesting and complex shapes, it is deemed easier for us to gain a feeling for surfaces by presenting them as natural extensions of curves. This approach occupies most of this chapter. Furthermore, it is often natural, convenient, and useful to study surfaces along a single profile which is the intersection of the surface with a plane (usually a vertical plane). The surface cuts this plane as a two-dimensional curve. Thus, a surface is simply a continuum of curves whose form can be more easily studied.

It is important to point out that the visual representation of surfaces involves considerably more computation than that of curves. Consider a curve $y = f(x)$ computed using n points of the independent variable x. If a surface $z = f(x, y)$ is to be represented with similar resolution, then n points in each direction (x and y) are required for a total of n^2 points. This can be a severe limitation on the speed of displaying surfaces when n becomes large. Moreover, the three-dimensional surface must always be projected to the two-dimensional plane of the viewing device, and this requires further calculations. Thus, in design work with surfaces, we will probably be more careful in selecting the right surface and less willing to cycle through many variations to arrive at the desired shape.

16.2. THREE-DIMENSIONAL SYMMETRY

We discussed the subject of symmetry earlier in relation to two-dimensional and three-dimensional curves. Symmetry of curves in a plane is usually examined in relation to one of the two axes although any arbitrary line may be an axis of symmetry for an appropriate curve. If an axis is a line about which the curve reflects, then the curve is symmetric about that axis. If an axis is a line about which the curve reflects and is, in additon, negated, then the curve is antisymmetric about that axis. The former is often referred to as even symmetry while the latter is often referred to as odd symmetry.

Considerable complexity is added to the symmetry question when curves become surfaces. There are now three principal axes and three principal planes each of which share two of these three axes in the Cartesian coordinate system. Symmetry of surfaces is usually discussed in terms of these planes (x-y, y-z, and z-x) and also in terms of the origin of the axes. (Note that any arbitrary plane may be a plane of symmetry for a surface and that, by suitable rotation of the surface, we can force that arbitrary plane to coincide with one of the principal planes.)

A surface is symmetric with respect to a plane if all the points of the surface on one side of the plane are reflected through the plane to form the other half of the surface. For instance, given $z = f(x, y)$, symmetry is represented functionally by

$$f(-x, y) = f(x, y) \text{ (symmetry about the } y\text{-}z \text{ plane)}$$
$$f(x, -y) = f(x, y) \text{ (symmetry about the } x\text{-}z \text{ plane)}$$

If we can express the surface as $x = f(y, z)$ or $y = f(z, x)$, then similar symmetry relations can be written for either of these expressions. It is possible for a surface to be symmetric about one, two, or all three planes; and we will give an example of each case.

A surface showing symmetry about the x-z plane only is $z = y^2/2 + x$ plotted in Figure 16.2.1. The viewpoint is nearly on the x-z plane of symmetry and somewhat above the $z = 0$ plane.

A surface showing symmetry in both the x-z and y-z planes is the paraboloid $z = x^2 + y^2$ plotted in Figure 16.2.2. The viewpoint is halfway between the negative x and negative y axes and somewhat above the $z = 0$ plane. This figure would appear identical from any other of the three viewpoints between x and y axes.

A surface showing complete symmetry for all three principal planes is the ellipsoid $z^2 = 1 - x^2/4 - y^2$ is plotted in Figure 16.2.3. The viewpoint is again halfway between the negative x and negative y axes and somewhat above the $z = 0$ plane.

When a surface has three planes of symmetry, it necessarily follows that it has *point symmetry*. This means that any point on the surface has a counterpoint on the line drawn from the point through the origin. All three points are colinear, and the two surface points are equidistant from the origin. We can rotate point-symmetric surfaces by π in any of the three principal planes without changing their appearance. If the surface is expressed as $z = f(x, y)$, then we write point symmetry as

$$f(-x, -y) = -f(x, y)$$

Another possible type of symmetry for surfaces is *axial symmetry*. In this case, any plane which contains the central axis of the figure will show the same profile of the

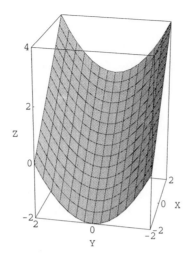

FIGURE 16.2.1. Surface for $z = y^2/2 + x$.

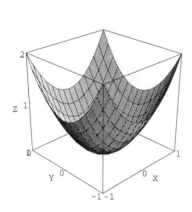

FIGURE 16.2.2. Surface for $z = x^2 + y^2$.

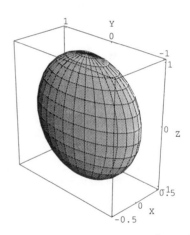

FIGURE 16.2.3. Surface for $z = 1 - x^2/4 - y^2$.

surface; also any plane which is normal to this axis will be intersected by the surface in a circle centered on the axis. In other words, the surface is invariant under a rotation about the central axis. Let $r = (x^2 + y^2)^{1/2}$ and $\theta = \arctan(y/x)$. Then we can write axial symmetry as

$$f(r, \theta) = f(r)$$

meaning that there is no dependence on the angle θ.

A surface is antisymmetric with respect to a plane when, in addition to being reflected through that plane, the points are reflected through one of the remaining planes. In the case where the surface is expressed as $z = f(x, y)$, antisymmetry would

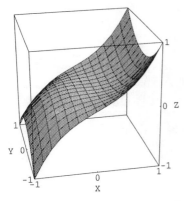

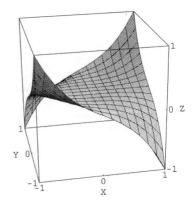

FIGURE 16.2.4. Surface for $z = x^3/2 + xy^2/2$.   **FIGURE 16.2.5.** Surface for $z = x^3y/2 + xy^3/2$.

be represented functionally by

$$f(-x, y) = -f(x, y) \text{ (antisymmetry about the } y\text{-}z \text{ plane)}$$
$$f(x, -y) = -f(x, y) \text{ (antisymmetry about the } x\text{-}z \text{ plane)}$$

Similar relations hold for when the surface is expressed as $x = f(y, z)$ or $y = f(x, z)$. A surface that illustrates antisymmetry about the y-z plane is $z = x^3/2 + xy^2/2$. It is plotted in Figure 16.2.4 with a viewpoint nearly along the y-z plane of antisymmetry looking from negative y. Note that the x-y plane also happens to be a plane of antisymmetry and that the x-z plane is a plane of symmetry.

A surface that is antisymmetric in both the x-z and y-z planes is $z = x^3y/2 + xy^3/2$ plotted in Figure 16.2.5. The viewpoint is nearly identical to that of Figure 16.2.4. Note that the surface also happens to be antisymmetric about the x-y plane; thus, it is antisymmetric in all three principal planes.

Other combinations of symmetry and antisymmetry are possible. Because symmetry (or antisymmetry) is often a desirable property of surfaces used in design, we will discuss how this is achieved in the various methods for forming surfaces in the following sections.

16.3. SURFACES FORMED BY TRANSLATING A CURVE

The simplest use of curves to create surfaces is to move the curve through space, without rotation, such that the trace of any point on the curve forms a space curve of movement. (Chapter 15 discussed space curves, which are merely curves in three dimensions.) The curve itself is often called the *generator* while the space curve along which it is translated is called the *directrix*. Note that, at any point of the generation of the surface, the generator is always parallel to its starting orientation. The space curve may have various levels of complexity, thus affecting the complexity of the resulting surface. First-level complexity is when the space curve is a straight line of arbitrary orientation. Second-level complexity is when the space curve is contained within a single plane again of arbitrary orientation. Third-level complexity is when

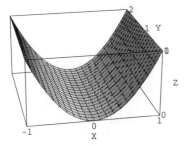

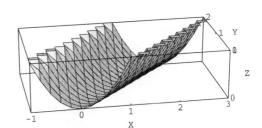

FIGURE 16.3.1. Surface for $z = x^2$.   **FIGURE 16.3.2.** Surface for $z = (x - y)^2$.

the space curve occupies a volume. We will examine these levels of complexity in turn.

If the two-dimensional curve is moved along a straight line perpendicular to its plane in order to form the surface, then whatever symmetry was associated with the original curve will be present in the surface. If the line of movement is oblique to its plane, then the symmetry is destroyed except in the special case that the line of movement is still in the plane that bisects the curve at right angles into its two parts. Let us illustrate these cases using the two-dimensional parabolic curve $z = x^2$. Let this curve be originally in the $y = 0$ plane. First move it along the $+y$ axis to obtain the surface shown in Figure 16.3.1. Note that the symmetry about the z axis of the original curve is continued into symmetry about the y-z plane for the surface. The viewpoint here is just off this plane of symmetry and somewhat above the $z = 0$ plane.

Figure 16.3.2 shows the result when we move the parabolic curve along a line parallel to the $z = 0$ plane but at an angle of $\pi/4$ between the positive x axis and the positive y axis. Now the y-z symmetry is gone, but the surface is symmetric about the vertical plane at an angle of $\pi/4$. However, if we take the line of movement to satisfy the requirement of the special case mentioned above by making it at an angle of $\pi/4$ between the positive z axis and the positive y axis, the surface shown in Figure 16.3.3 is produced. Note that, again, the y-z plane is a plane of symmetry for the surface.

Now we consider the second level of complexity: the line of movement is not straight but is contained in one plane arbitrarily oriented in space. In this case, it is not possible to retain the symmetry of the original curve throughout the generated surface except in the special case when the plane of the curve movement is that which bisects the original curve at right angles. In this case, we have a generalization of the kind shown for straight-line movement in Figure 16.3.3.

However, if the line of movement is itself a symmetric curve within its own plane and its line of symmetry is parallel to the plane of the original curve, the generated surface will be symmetric about the plane which contains this line of symmetry. An example should help to clarify this. Consider the same parabolic two-dimensional curve in the x-z plane as used in Figure 16.3.1. Now move its apex point along $z = y^4$, which is symmetric in the y-z plane, from $y = -1$ to $y = +1$ to generate the surface shown in Figure 16.3.4. This surface is symmetric about both the x-z and y-z axes. We should be able to discern that, if the plane of the line of movement were tilted rather than vertical, the symmetry about the y-z plane is destroyed.

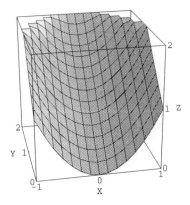
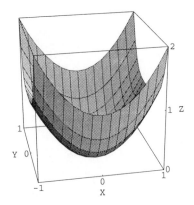

FIGURE 16.3.3. Surface for $z = x^2 + y$.   **FIGURE 16.3.4.** Surface for $z = x^2 + y^4$.

In the case of third-level complexity, the line of movement is a space curve and no symmetry can occur in the generated surface, regardless of the symmetry of the original curve. For this reason, the generated surfaces can have a wide variety of interesting, but perhaps not esthetic, shapes. We give no example of this type of surface.

16.4. SURFACES OF REVOLUTION

In contrast to Section 16.3, if a curve defined for a positive independent variable, $z = f(x)$ or $z = f(y)$, is rotated a full turn about the axis of the dependent variable, then a *surface of revolution* is generated. If the curve is symmetric and defined for equal lengths of the positive and negative axes of the independent variable, then only a half turn about the z axis is necessary. If the curve is antisymmetric, then a full turn is necessary; and the resultant surface will be symmetric about the x-y plane. Let the axis about which the curve turns be termed the rotation axis (taken to be the z axis). For n intersections of the rotation axis by the curve, assumed to be continuous, there will be $n - 1$ separate volumes enclosed by the generated curve. Surfaces of revolution have, by definition, axial symmetry about this rotation axis.

Let us show a simple example of such a surface. Consider the curve $z = \exp(-x^2)$ defined on the positive x axis (see Figure 16.4.1). A full turn of the curve about the z axis generates the surface $z = \exp(-x^2 - y^2)$, as shown in Figure 16.4.2 out to a radius of two. For this surface, the defining curve intersected the rotation axis just once; and so there is no enclosed volume. The plot shows two sets of intersecting lines that are important in characterizing the surface. Those lines that emanate from the axis of rotation and are aligned in a vertical plane are called the *meridians*, and those lines that are normal to them and which are circles at constant z are called the *parallels*. Any meridian is, of course, identical in shape to the generating curve.

Now consider the curve $x = \cos(2z/\pi)$ defined for $-1 \leq z \leq +1$ as seen in Figure 16.4.3. This curve intersects the z axis at -1 and $+1$. If it is rotated a full turn about the z axis, the surface in Figure 16.4.4 is generated. This surface now encloses a

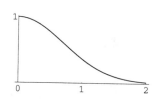

FIGURE 16.4.1. Graph of exp($-x^2$).

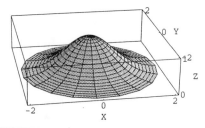

FIGURE 16.4.2. Surface for $z = \exp(-x^2 - y^2)$.

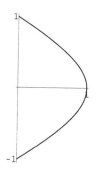

FIGURE 16.4.3. Graph of $x = \cos(2z/\pi)$.

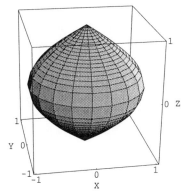

FIGURE 16.4.4. Surface for $z = \arccos[(x^2 + y^2)^{1/2}]/(\pi/2)$.

single volume. Additional volumes would have been enclosed if the original curve had been defined for a larger range of z that included more intersections.

For surfaces of revolution about the z axis, the dependence of z on x in the two-dimensional curve is replaced by a dependence on $(x^2 + y^2)^{1/2}$. More simply, using $r = (x^2 + y^2)^{1/2}$, we can write $z = f(r)$ or $r = f(z)$. It is clear that any vertical plane that includes the z axis is a plane of symmetry. However, symmetry about the x-y plane only exists if the original curve $z = f(x)$, or $x = f(z)$, is symmetric about $z = 0$.

What if the curve is antisymmetric about the z axis? Then the surface of revolution will be constricted to the origin, will be symmetric with respect to the x-y plane, and will again enclose $n - 1$ volumes for every n intersections of the z axis by the original curve. If the curve is antisymmetric but only has the one intersection at $z = 0$, then the generated surface has two parts connected only at the origin. Such a surface is generated by the curve $x = \sin(2z/\pi)$, defined for $-1 \leq z \leq +1$, as shown in Figure 16.4.5. The surface generated by the revolution of this curve is shown in Figure 16.4.6.

Many familiar shapes are simply surfaces of revolution. Using a circle of radius 1/2 as a generating curve, let us displace it from the origin to have a center at (1, 0). The result is shown in Figure 16.4.7. The figure generated by rotating this circle about the z axis is called the *torus* and is shown in Figure 16.4.8.

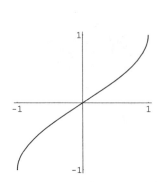

FIGURE 16.4.5. Graph of $x = \sin(2z/\pi)$.

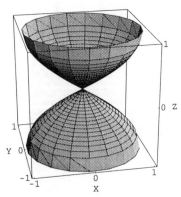

FIGURE 16.4.6. Surface for $z = \arcsin[(x^2 + y^2)^{1/2}]/(\pi/2)$.

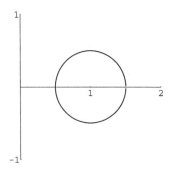

FIGURE 16.4.7. Circle translated to $(1, 0)$.

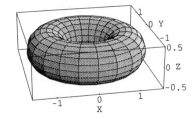

FIGURE 16.4.8. Torus generated from circle.

16.5. RULED SURFACES

16.5.1. Ruled Surfaces by Translation

Ruled surfaces are a special class of surfaces generated by the motion of a straight line. This motion can be translation or rotation or a combination of both. The variety of surfaces that can be generated by a straight line is surprisingly large, including very basic shapes such as cylinders and ranging up to very complex shapes. We often refer to the line used to generate the surface as the generator. It may be finite or infinite in length, and the generated surfaces are correspondingly bounded or unbounded.

Those surfaces generated by a simple translation of the generator will be considered first. The very simplest would be the plane generated by translating a straight line along a line perpendicular to it. Another surface would be the cylinder, generated by moving a line about an axial line parallel to it in a circular or other closed translation path. A representation of this surface (circular cylinder) is shown in Figure 16.5.1 with the generator plotted at intervals of five degrees. A "belt" has been added to the middle of the cylinder.

The cylinder becomes distorted when the generator is tilted relative to the axial line by rotating the generator at the belt-line. The line always lies in a plane tangent to the

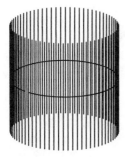

FIGURE 16.5.1. Cylinder as a ruled surface.   **FIGURE 16.5.2.** Hyperboloid as a ruled surface.

cylinder. The type of surface which will be generated in this way is the *hyperboloid of one sheet*, which is given by the equation

$$z = c(x^2/a^2 + y^2/b^2 - 1)^{1/2}$$

One instance of this type of ruled surface is shown in Figure 16.5.2. Note that the belt now surrounds the smallest circumference of the hyperboloid; this region is referred to as the *stricture* of the ruled surface. The cone is the limit of the hyperboloid as the belt tightens to zero radius. Its functional form is

$$z = c(x^2/a^2 + y^2/b^2)^{1/2}$$

The example of a cone in Figure 16.5.3 uses $a = b = c = 1$.

16.5.2. Ruled Surfaces by Rotation

Other ruled surfaces may be generated by rotating the generator about a fixed point. A simple surface generated by rotation of a line is the flat disk. Also, the cone shown in Figure 16.5.3 may be considered to be generated in this way if the rotating line is tilted in relation to the rotation axis. More complex surfaces can be created by making the rotating line pass through previous positions of the line. For instance, a "double" cone can be created, as shown in Figure 16.5.4. In this figure, the radial distance of the generator from the central axis takes the form $r = c\cos^2(\theta)$ where c is proportional to z.

Many different surfaces can be generated in this manner by letting $r = f(\theta)$ and rotating the generator accordingly. We can generate a different set of surfaces by having simultaneous rotation in the plane perpendicular to the rotation axis and oscillation in the vertical plane. For instance, consider the generator that rotates in the x-y plane to create a flat disk. Now impose a harmonic oscillation of the z position at the tips of the generator as it rotates. If the frequency of the oscillation is n times the rotation rate in the x-y plane, n lobes will be created on the disk. The number n must be odd in order that the generator be a straight line. Figure 16.5.5 shows how this ruled surface appears when $n = 5$.

FIGURE 16.5.3. Cone as a ruled surface.

FIGURE 16.5.4. Double cone.

FIGURE 16.5.5. Fluted disk.

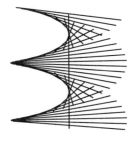

FIGURE 16.5.6. Complete helicoid.

16.5.3. Ruled Surfaces by Translation and Rotation

If we allow the straight-line generator to undergo both translation and rotation, the variety of surfaces that can be generated is fairly unlimited. Only a few examples will be given here. One of the most common surfaces that can be generated in this way is the helicoid, for which the fundamental example may be the twisted ribbon. This is shown in the Figure 16.5.6. The central axis of the helicoid has been drawn to help illustrate it. All lines are centered on this axis and are normal to it as they rotate about it with increasing upward translation. This figure is labeled as the complete helicoid because the lines extend on both sides of the axis to equal distances.

We can generate another realization of the helicoid by letting the lines occupy only one side of the central axis as shown in the Figure 16.5.7. This partial helicoid can be further restricted with a shorter rotating line displaced from the central axis by a constant amount. Figure 16.5.8 shows the helicoid when the line is displaced from the central axis by an amount equal to its length. Again, the helicoid is generated by rotating this line as before while translating it upward. The result is what might be viewed as a spiral staircase.

We can make another example of a ruled surface by moving a straight line along

FIGURE 16.5.7. Partial helicoid.   **FIGURE 16.5.8.** Partial helicoid.

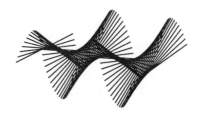

FIGURE 16.5.9. Reciprocating wave surface.

a second straight line perpendicular to it but now oscillating the line up and down in the plane perpendicular to this second line. Let the oscillating motion be simply the sine function. Figure 16.5.9 shows the result, and it appears to be related closely to the helicoid.

We should recognize that these ruled surfaces can always be described functionally with $z = f(x, y)$ or parametrically with the equations given in Section 16.1. However, it is useful to see how they arise from straight-line generators. Because they arise from such generators, ruled surfaces can always be cut by certain planes that will show straight lines as the intersection of the surface with the cutting plane.

16.6. SURFACES PRODUCED BY FUNCTIONS

16.6.1. Polynomials

Up to this point, surfaces in this chapter have been produced by the motion of lines that are either straight or are some functional form in themselves. This section covers, with a few examples, some of the basics of surfaces represented by simply $z = f(x, y)$ or by the parametric form $x = f(u, v);\ y = g(u, v); z = h(u, v)$. The functional form is related to visual features of the surfaces that are important in a design context. Recall that $y = f(x)$ was the functional expression for curves on a plane. Many of the visual features of plane curves can be extrapolated to surfaces by comparing the functional form of surfaces with those forms already familiar from the study of plane curves. Symmetry was already discussed in Section 16.2 because of its unique importance with surfaces in design contexts.

When the expression for surfaces is given by a polynomial of k terms

$$z = \sum_{i=1}^{k} a_i x^n y^m$$

with n and m arbitrary or by a rational polynomial with k terms in the numerator and l terms in the denominator

$$z = \sum_{i=1}^{k} a_i x^n y^m \bigg/ \sum_{j=1}^{l} b_j x^n y^m$$

then many of the attributes of plane curves, as presented in Chapter 2, apply to these surfaces. The visual effects that can be controlled by the form of the equations for plane curves can also be controlled by suitable adjustment of the polynomial form of a surface.

In particular, Chapter 2 showed how the zeroes and poles of the plane curve were controlled by the numerator and denominator, respectively, of the rational polynomial. These zeroes or poles were points on the x axis at which $y = 0$ or ∞. For surfaces, the zeroes of $z = f(x, y)$ are where the surface intersects the x-y plane; this intersection will be defined by a line or set of lines rather than a point or set of points. The line (or lines) will have a functional form in itself and may be open or closed in the x-y plane. In certain cases the line (or lines) may collapse to a single point (or points). The poles of $z = f(x, y)$ are where the denominator becomes zero and the surface then goes to infinity. Analogous to zeroes, the poles may be lines or points of the surface. For instance, consider the surface

$$z = \frac{x^2 + y^2 - \frac{1}{4}}{4(x^2 - 1)}$$

This surface has zeroes where $x^2 + y^2 - 1/4$ equals zero. This is the circle on the x-y plane of radius 1/2; therefore, the surface must intersect the x-y plane along this circular ring. The poles of the surface are where $x^2 - 1$ is equal to zero or at $x = \pm 1$; these poles will occupy two straight lines along these two x positions. The surface is shown in Figure 16.6.1. Note that in this case, the pole lines are where the surface goes from negative infinity on one side to positive infinity on the other.

As the complexity of the rational polynomial increases, the surfaces given by $z = f(x, y)$ will become very complex; and prediction of their visual appearance becomes difficult or impossible. It is convenient to analyze their appearance along the x and y axes and also along any arbitrary line between the axes. Equations for the slices of the surface in the vertical planes containing these lines can be easily derived from the full expression by setting $x = 0$ or $y = 0$ to get the slices along the axes and $x = cy$, with c a constant, to get the slice along any oblique line. For the example surface in Figure 16.6.1, these three equations are

$$z = \frac{\frac{1}{4} - y^2}{4} \quad (x = 0)$$

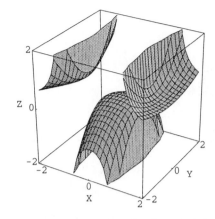

FIGURE 16.6.1. Surface for $z = [(x^2 + y^2 - 1/4)/(x^2 - 1)]/4$.

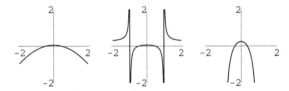

FIGURE 16.6.2. Vertical slices of $z = [(x^2 + y^2 - 1/4)/(x^2 - 1)]/4$ (left: y-z plane; middle: x-z plane; right: between x and y axes).

$$z = \frac{x^2 - \frac{1}{4}}{4(x^2 - 1)} \quad (y = 0)$$

$$z = \frac{(1 + c^2)y^2 - \frac{1}{4}}{4(c^2 y^2 - 1)} \quad (x = cy)$$

Graphs of these three functions are given in Figure 16.6.2; the third equation uses $c = 1$; therefore, the graph is the profile of the surface along the diagonal $x = y$. In designing surfaces for specific applications, it should be efficient and beneficial to examine the behavior of these profiles in relation to the zeroes and poles of the function rather than to generate the full surface for many iterations.

A special surface is the surface of revolution discussed in Section 16.4. For such surfaces, the profile is identical no matter which plane is used, provided that the plane passes through the origin. Functionally, these surfaces are given by $z = f(r)$ where $r = (x^2 + y^2)^{1/2}$. Conversely, whenever the functional form $z = f(x, y)$ can be reduced to $f(r)$, the surface must have the properties of a surface of revolution. Thus, any zeroes or poles of such functions must occur as circular rings in the x-y plane; in the limit, this may reduce to the single point at the origin. A slight modification of the surface of Figure 16.6.1 will produce a surface of revolution. Consider the equation

$$z = \frac{x^2 + y^2 - \frac{1}{4}}{x^2 + y^2 - 1}$$

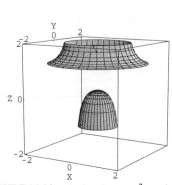

FIGURE 16.6.3. Surface for $z = (r^2 - 1/4)/(r^2 - 1)$.

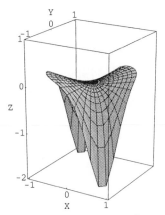

FIGURE 16.6.4. Surface for $z = 2r^2 \sin(2\theta)/[2 + r^2 \sin(2\theta)]^2$.

which reduces to

$$z = \frac{r^2 - \frac{1}{4}}{r^2 - 1}$$

The zeroes occur on the circle of radius 1/2 in the *x-y* plane while the poles occur on the circle of radius unity in the same plane. The plot of this function appears in Figure 16.6.3. This surface is in fact two surfaces, separated by the circle of poles at $r = 1$.

Another special type of surface arises when the polynomials comprising the surface's expression consist only of terms having $x^n y^n$ in them, with the exponents equal (including $n = 0$ to give constants). Recall that, in polar coordinates, $x = r\cos(\theta)$ and $y = r\sin(\theta)$. Thus, the quantity xy is equal to $r^2 \cos(\theta) \sin(\theta)$ which, using $\sin(2\theta) = 2\cos(\theta)\sin(\theta)$, can be reduced to $r^2 \sin(2\theta)/2$. If the surface $z = f(x, y)$ is equivalent to $z = f[r^2 \sin(2\theta)]$, it has a predictable behavior. Because of the $\sin(2\theta)$ dependence, it must be periodic in θ with period π. Being so, the surface must have two planes of symmetry: in analogy to plane curves having two lines of symmetry when they are circularly periodic with period π. The planes of symmetry are along the two diagonals given by $x = y$ and $x = -y$, which are at $\theta = \pi/4$ and $\theta = 3\pi/4$. This follows from the fact that $\sin[2(\pi/4 + \alpha)] = \sin[2(\pi/4 - \alpha)]$ for any α, and similarly for $3\pi/4$. As an example, consider the following equation:

$$z = \frac{xy}{1 + 2xy + x^2 y^2}$$

Using the *x* and *y* substitutions in terms of *r* and θ, this reduces to

$$z = \frac{2r^2 \sin(2\theta)}{[2 + r^2 \sin(2\theta)]^2}$$

This surface is shown in Figure 16.6.4 for $0 < r < 1$.

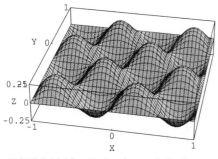

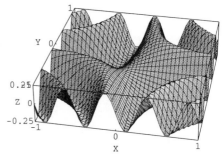

FIGURE 16.6.5. Surface for $z = \sin(2\pi x)$
$\sin(2\pi y)/4$.

FIGURE 16.6.6. Surface for $z = \sin(4\pi xy)/4$.

16.6.2. Transcendentals

In extrapolating from curves to surfaces that contain transcendental functions (such as sin, exp, log), there are a few concepts that are helpful in predicting the shape of the surfaces. The first concept is that of orthogonality of factors. The expression for a surface, given by $z = f(x, y)$, has orthogonal factors when it can be separated into two or more factors, none of which contains both x and y. For instance, the following are orthogonal:

$$\sin(x)\cos(y)$$
$$\cos(x)\exp(y^2)$$
$$\sin(4x)\exp(1 - x)/\tan(y)$$

But these expressions are not orthogonal:

$$\cos(x + y)\exp(-2x)$$
$$\cos(x)\exp[-(x^2 + y^2)]$$
$$\log(|xy|)\sin(2\pi y)$$

The factors need not be composed of transcendentals only but may include polynomials. If the expression is orthogonal, the surface consists of the product of two surfaces each invariant in one or the other of the two directions x and y. Thus, we can visualize the shape of the surface if we can visualize the shape of the two orthogonal surfaces by themselves. For example, consider the orthogonal surface $z = \sin(2\pi x)\sin(2\pi y)/4$. We have a wave in one direction multiplied by a wave in the other direction to give the surface shown in Figure 16.6.5.

Another surface that is relatively easy to analyze is generated by transcendental functions whose arguments have terms all in the form $x^n y^n$ (including $n = 0$ to give constants). This situation is analogous to the one discussed in Section 16.6.1 with respect to polynomials. Again the result is that xy can be replaced everywhere by $r^2 \sin(2\theta)$ and that the surface will consequently be periodic with period π in the x-y plane. For instance, consider the surface given by $z = \sin(4\pi xy)/4$ plotted in Figure 16.6.6. Note that the planes of symmetry contain the diagonals along $x = y$ and

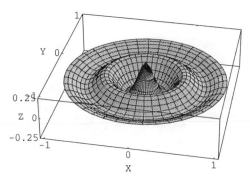

FIGURE 16.6.7. Surface for $z = \cos(4\pi r)e^{-2r}/4$.

$x = -y$; we discussed this feature for surfaces formed from polynomials having the same property.

A third type of surface that may be considered a special class is one in which all the arguments of the transcendental functions are in the form $(x^2 + y^2)^{p/2}$ where the exponent $p/2$ is arbitrary. This form can be replaced everywhere by simply r^p; and the surface, expressed as $z = f(r)$, is then necessarily axisymmetric or a surface of revolution. This shape was treated in Section 16.4. As an example using transcendental functions, consider $z = \cos(4\pi r)\exp(-2r)/4$ plotted in Figure 16.6.7.

Transcendentals alone can generate surfaces of considerable complexity, especially when none of the simplifying concepts discussed here applies. By combining transcendental functions and polynomials, surfaces can become even more complex and difficult to visualize without actually producing a plot. Again, we can gain some insight by often plotting the plane curves contained in the intersections of the surface with the *x-z* and *y-z* planes or with any arbitrary vertical plane. It may be judicious to work with such "cross-sections" before finally selecting a functional representation that will satisfy one's design goals.

16.7. SUMMARY

Surfaces can be considered as extensions of two-dimensional curves. By translating and rotating two-dimensional curves through space, three-dimensional surfaces are formed. Many of the surfaces in common use can be formed in this way. Surfaces exhibit the same properties assigned to two-dimensional curves: for instance, symmetry, continuity, poles and zeroes, and periodicity. The difference is always that these are more difficult to visualize and adequately represent in plotted figures because of the need to project the three-dimensional figure to a two-dimensional viewing plane.

REFERENCES

1. Farin, G., *Curves and Surfaces for Computer Aided Geometric Design: A Practical Guide*, Academic Press, 1988.

2. Mortenson, M. E., *Geometric Modeling*, John Wiley & Sons, 1985.

3. von Seggern, D. H., *CRC Handbook of Mathematical Curves and Surfaces*, CRC Press, 1990.

4. von Seggern, D. H., *CRC Standard Curves and Surfaces*, CRC Press, 1993.

INDEX

A

affine transformation, 157
all-pass filter, 217
amplitude spectrum, 207
angular scaling, 165
antisymmetric functions, 8
antisymmetry, 17, 208
approximating functions, 184
axes, 1
axial symmetry, 248

B

B-spline, 198
bandpass filter, 217
basis functions, 181
bell curve, 46
bending transformation, 174
Bernstein polynomials, 186
Bezier curves, 186
binomial factors, 9
Bowditch curves, 129
breakpoints, 197

C

Cartesian coordinates, 1
circular helix, 237
closed curve, 5, 123
coefficients, 7
column vector, 155
compact basis, 181
compound cycloid, 64
compound harmonic, 55
compound transform, 157
compression, 167
cone, 255
conformal transform, 156
conic sections, 30, 41
constants, 4
continuity for curves, 193
continuous curve, 3
continuous function, 193
control points, 187
convex polygon, 188
coordinate pair, 1
coordinate system, 1
cosine, 53
critical points, 10

cubic, 7
cubic bending, 173
curtate cycloid, 61
cycloids, 61
cylindrical coordinates, 235

D

dependent variable, 4
derivative, 10, 45
dilation factor, 225
directrix, 250
discontinuity, 20, 193, 209
discontinuous approximations, 193
discrete Fourier transform, 208
domain, 4
dot product, 155
double helix, 241

E

ellipses, 30, 36
ellipsoid, 248
epicycloid, 88
epitrochoids, 86, 127
essential discontinuities, 194
even functions, 8
even harmonic, 54
expansion, 167
explicit form, 5
exponent, 7
exponential decay, 45
exponential function, 45, 150
exponential growth, 45
exponential ramp, 49

F

factor, 7
filter, 217
fixed points, 156
Fourier series
    amplitude spectrum, 207
    analysis, 207, 208
    coefficients, 207
    discrete transform, 208
    fundamental frequency, 205
    fundamental period, 53, 205
    orthogonal basis, 207
    phase spectrum, 207